# Rural Development

# Rural Development

## Principles and Practice

**Malcolm J. Moseley**

SAGE Publications
London • Thousand Oaks • New Delhi

First published 2003

SAGE Publications Ltd
6 Bonhill Street
London EC2A 4PU

SAGE Publications Inc
2455 Teller Road
Thousand Oaks, California 91320

SAGE Publications India Pvt Ltd
B-42, Panchsheel Enclave
Post Box 4109
New Delhi 110 017

**British Library Cataloguing in Publication data**

A catalogue record for this book is available from the British Library.

ISBN 0 7619 4766 3
ISBN 0 7619 4767 1 (pbk)

**Library of Congress Control Number: 2002108292**

Typeset by C&M Digitals (P) Ltd., Chennai, India
Printed in India by Gopsons Papers Ltd., Noida

## Contents

# Preface

This book aims to distil much of what I have learned in the past 15 years or so about the local promotion of rural development in Britain, Ireland and continental Europe. In that regard I was fortunate to be, from 1987 to 1993, the first director of ACRE, the national voluntary organisation committed to promoting the vitality of England's villages and small towns and to improving the quality of life of their more disadvantaged residents. And from 1993, to the present day, I have been equally fortunate to work as a researcher, teacher and consultant in the Countryside and Community Research Unit of what is now England's newest university, the University of Gloucestershire.

During those 15 years, frequent contact with policy-makers and practitioners engaged in rural development, with unpaid activists working at the local level and with a variety of students, some of them already with a foot in the world of practice, persuaded me of the need for a concise text on the challenge of undertaking locally focused rural development. Hence this attempt to draw together a mixture of evidence and opinion around a number of core themes or issues, one per chapter, which together embrace much of the substance of that challenge.

In writing, I have borne in mind four types of potential reader. The first is undergraduate and master's-level students taking courses in 'rural something', for example rural geography, sociology, economics or planning, and for whom some understanding of local development is important. Also relevant in academia are research students, and researchers more generally, coming into 'rural development' from more specialist backgrounds. The second type is a range of local or national activists keen to improve the well-being of our rural communities – for example parish clerks, local councillors, members of local voluntary bodies and of local and national amenity or social welfare organisations. Third are those salaried practitioners who find themselves engaged in some aspect of rural development despite having received little or no formal training in the subject. Such people generally work in local or central government and the various development agencies and partnerships. Fourth is a variety of specialists in related professions and disciplines, such as community workers, conservation officers, agriculturists and land-use planners, who want to learn more about a related endeavour.

As for the approach, each chapter attempts to link theory and practice, giving roughly equal weight to each. 'Theory' because it seeks to structure and make sense of the mass of seemingly unconnected facts to which we are otherwise confined; 'practice' because it serves to ground theory in the muddy and murky world of real-world struggles to get things done. It is theory and practice taken together that best meets the needs of students and practitioners of rural development, who each tend in my experience to have a commendable aversion both to 'theory for theory's sake' and to the indiscriminate accumulation of facts, case studies and examples of good practice.

Next, a confession. About one third of the 200 or so references to literature cited in the text are books, papers or reports written either by myself – often with colleagues – or by/for the two rural development agencies with which I have had most dealings over the past decade. These are the Brussels-based LEADER Observatory and England's Rural Development Commission (plus the Countryside Agency, into which the latter was subsumed in 1999). This selectivity reflects more my goal of drawing substantially on personal experience than any suggestion that those sources contain a disproportionate share of what is worth knowing on the subject.

As for the content of the book, an introductory chapter sets out the underlying argument and structure and is followed by 13 substantive chapters which fall into two groups. First are nine devoted to overarching concepts or themes in rural development. Not *the*, much less *the only*, concepts or themes but nine which after some thought seem to capture most of what is significant. Certainly sustainability, innovation, adding value, entrepreneurship, community, social inclusion, accessibility, partnership and community involvement all occur time and again in the recent literature on rural and local development. It may well be that further chapters, perhaps on capacity building, networking, integration and governance, to name just four other contenders for inclusion, would also have been appropriate. But each is covered to some extent in one or other of the nine thematic chapters.

The remaining four chapters are devoted to particular aspects of the systematic pursuit at a local level of those overarching themes. They relate to the diagnosis of a local area, strategic planning, implementation and evaluation. But the distinction between the two groups of chapters is not clear-cut. Certainly many of the themes of the first group of nine chapters are effectively means, as well as ends, of local rural development. Indeed, the fact that 'product' and 'process' are inherently intertwined and sometimes interchangeable is one of the great lessons of local development repeated time and again in this book.

All 13 chapters follow a common format. The first part is a concise statement of the particular concept's significance in a rural development context, and a suggested definition of it. The second is a brief consideration of some key issues surrounding it, again from a rural development perspective. The third is 'the toolkit' – a listing and brief critique of a number of ways of practically pursuing or undertaking the concept or task in the real world.

After that come two case studies per chapter, 28 in all, collectively comprising about one third of the book. Each has been chosen and written to illustrate key issues raised in the immediately preceding text and to help link theory and practice. Seventeen of the 28 present work in which I was involved as a practitioner, researcher or consultant and a further six relate to programmes or initiatives of which I have personal knowledge. The other five have been condensed from the literature to best illustrate some aspect of practice outside my direct experience. Twenty-one of the total relate exclusively to some part or parts of the UK and the other seven to places in one or more of the other member states of the European Union (see the map which follows).

While responsibility for what follows is exclusively mine, many people have kindly helped as collaborators in the various studies which have underlain this book, or as constructive critics of draft chapters. Particular thanks in this respect are due to: Phil Allies, Joan Asby, David Atkinson, Madeline Barden, Sally Bex, Ros Boase, Sian Brace, Yves Champetier, Catherine Chater, Mike Clark, Wendy Cutts, James Derounian, Michael Dower, Ged Duncan, Pam Ellis, Anne Fromont, Laurie Howes, Tony Kerr, Malcolm Kimber, Catherine le Roy, Nick Mack, Ruth McShane, Stephen Owen, Michael Palmer, Gavin Parker, Ian Purdy, Carl Sanford, Lesley Savage, Paul Selman, Denise Servante, Elisabeth Skinner, Denise Sore, Monika Strell, Richard Tulloch, Andrew Wharton, Louise Wilby, Amanda Wragg and Stephen Wright.

More specifically, my grateful thanks go to two colleagues whose support deserves particular mention: Trevor Cherrett of Sussex Rural Community Council – fellow researcher on the PRIDE research project, co-author of the chapter on 'partnerships', which is based mainly on that work, and genial collaborator in much that I have done in recent years; and Nigel Curry, who as head of the Countryside and Community Research Unit at the University of Gloucestershire encouraged me to devote a good deal of 2000 and 2001 to researching and writing this book, while knowing full well that it would add nothing to the Unit's research income stream or to its standing in the cross-university 'Research Assessment Exercise'.

I also gladly pay tribute to the real heroes of rural development – the people who turn up on dark winter evenings to manage the village hall, drive the community minibus, plan the parish appraisal or organise the good neighbours scheme. Such stalwarts have provided much of the inspiration, and indirectly the material, for this book and I hope that in some small, albeit circuitous, way it will add strength to their elbows.

Finally, I dedicate this book with love to Helen. Laid low for so long by severe ME she could and can offer words of encouragement only from her bed. May we once more walk through the English countryside together.

Malcolm Moseley
Cheltenham

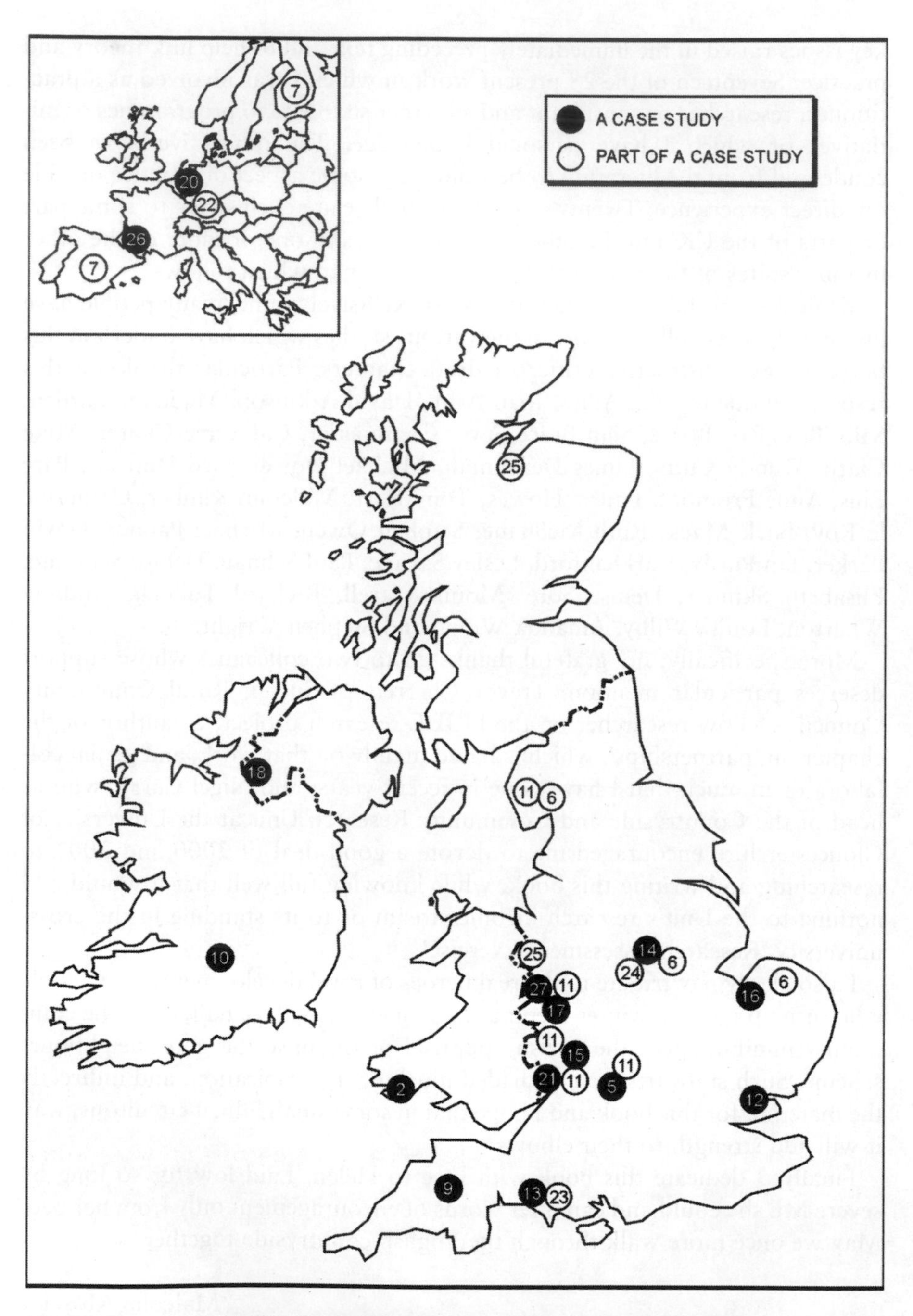


***The case studies in Britain and elsewhere in Europe. (See the Contents pages to identify the individual cases. Case Studies 1, 3, 4, 8, 19 and 28 are not indicated as they relate less directly to particular places.)***

# 1

# Rural Development: Making It Local

'You've got the crowd, you know the pitch...' (David Beckham on the advantages of playing at home)

This book is about 'rural development', about the attempts being made in Britain and other parts of Europe to address in a co-ordinated and locally sensitive way the range of pressing economic, social and environmental problems that beset the continent's rural areas. More specifically it is about some of the fundamental issues and concepts that underlie that intervention – concepts that relate both to the rationale of rural development and to the manner of its realisation.

Those issues and concepts form the focus of the 13 chapters that follow this introduction, but given the underlying thesis of the book, that rural development can only be pursued successfully at the local level, none of them is more important than *local development*, which involves bringing to bear the full range of local resources, human and material, to resolve identified concerns. The task of this first chapter then is to explore the meaning and purpose of 'rural' and 'local' development and to present as a case study the LEADER programme, both as a pan-European rural development programme devised in Brussels and as a local development venture carried out in a small part of rural Wales.

But first, the word 'rural'. A considerable literature exists on what 'rural' might mean and, indeed, on whether 'rurality' is really significant in the context of advanced western society in the late twentieth and early twenty-first centuries (see for example Denham and White, 1998; Dunn *et al.*, 1998; Shucksmith *et al.*, 1996). Here, however, we will be heavily pragmatic, simply defining 'rural areas' as those with 'low population density containing scattered dwellings, hamlets, villages and small towns', and effectively put to one side such questions as 'How low is "low"?' and 'How small is "small"?', since there is no agreed answer to such questions, the 'cut-off points' of density and settlement size being best set according to the task in hand.

The point is that an emphasis on population density – rather than on other possible criteria of rurality with strong competing claims such as land use, economic structure, culture and remoteness – usefully focuses attention on what, in the context of development initiatives, are three crucial elements of the rural scene:

- the fact that all rural people, and many of the economic, social, political and cultural activities which are relevant to their well-being, are by definition located in isolated buildings or in settlements that are both small and widely separated;
- the fact that the wide expanses of land that necessarily separate them are subject to a mass of powerful and competing demands and pressures as agriculture and other forms of land-extensive economic activity are compelled to restructure; and
- the fact that an increasingly prosperous and 'space hungry' urban population is drawn, in a variety of ways and for a variety of reasons, both to those small settlements and to the wide expanses of land that separate them.

That essential rural context has certainly conditioned, even if it has not 'caused', a set of inter-related concerns that have intensified in recent years and which underlie the various calls made for 'rural development' programmes. Those concerns are not universal – indeed we will later stress the diversity of rural Europe – but in varying ways and to a varying extent they are certainly widespread and keenly felt. The following is an indicative, and by no means exhaustive, list:

- First are some *economic concerns* which derive from the reduced and still reducing ability of land-extensive economic activities – notably agriculture, forestry, quarrying and mining – and of many other rural industries linked closely to them to provide secure employment and adequate incomes for the people engaged in them. Other 'economic' concerns relate not to the challenge of reformulating and complementing land-based industry but to the costs of servicing a widely scattered population that offers little in the way of economies of scale.
- Second are various *social and cultural concerns* which are often subsumed in the expression 'rural deprivation'. They include un- and under-employment, low incomes, social exclusion, insufficient affordable housing for local people, the steady decline of local services and facilities and a deeper cultural malaise linked to the erosion of caring local communities, a sense of powerlessness in the face of rapid change, and latent or overt conflict between long-established residents and many newcomers with different sets of values.
- Third are *environmental concerns* which stem particularly from agricultural intensification and a consequent decline in wildlife and in habitat and

countryside diversity. They derive also from the growing pressures placed on the countryside by an urban population that is increasingly keen to live, work and/or enjoy its leisure time there.

- Fourth, to these may be added some *political and institutional concerns* related to the lack or frequent inadequacy of the machinery necessary to resolve such concerns at the local level in a way that recognises both their inter-relatedness and the vital need for collaborative working between a host of agencies and actors including local residents themselves.

Sometimes such concerns are expressed indirectly in 'vision statements' which encapsulate what would prevail if they were satisfactorily resolved. A recent example is that of the British government set out in a 'Rural White Paper' summarising its policies for rural England (Department of Environment, Transport and the Regions and Ministry of Agriculture, Fisheries and Food 2000: 6):

> Our vision is of
>
> - a living countryside with thriving rural communities and access to high quality public services
> - a working countryside with a diverse economy giving high and stable levels of employment
> - a protected countryside in which the environment is sustained and enhanced and which all can enjoy
> - a vibrant countryside which can shape its own future...

Much the same sentiment had been expressed four years earlier in a declaration issued jointly by several hundred 'rural leaders' drawn from across Europe and meeting in Cork under the aegis of the European Union. The 'Cork declaration' of November 1996 (LEADER Observatory 1997a) marked a significant step on the road from narrow agricultural and other sectoral policies applied to rural Europe in general, towards specifically rural policies and programmes respecting the needs and resources of local areas. Its action plan made explicit the need for integrated rural development policy with a clear territorial dimension, the diversification of economic activity, respect for the tenets of sustainability and of subsidiarity (meaning the 'decentralisation' of decision-making) and improved mechanisms for planning, managing and financing rural development at the local level.

## RURAL DEVELOPMENT

This brings us to the definition of *rural development*. The following three suggested definitions build on the above brief discussion of 'rurality' and of associated concerns and aspirations to encapsulate what most contemporary commentators understand by the term:

- 'a broad notion encompassing all important issues pertinent to the collective vitality of rural people and places... [including] education, health, housing, public services and facilities, capacity for leadership and governance, and cultural heritage as well as sectoral and general economic issues...' (OECD, 1990: 23);
- 'a multi-dimensional process that seeks to integrate, in a sustainable manner, economic, socio-cultural and environmental objectives' (Kearney *et al.*, 1994: 128); and
- 'a *sustained* and *sustainable process* of *economic, social, cultural* and *environmental* change *designed* to enhance the *long-term well-being* of the *whole community*' (Moseley, 1996b: 20).

The third of these definitions includes 12 italicised words which are central to the understanding of 'rural development' and to its promotion:

- *sustained*...not short-lived;
- *sustainable*...respecting our inherited 'capital';
- *process*...a continuing and inter-related set of actions;
- *economic*...relating to the production, distribution and exchange of goods and services;
- *social*...relating to human relationships;
- *cultural*...relating to 'ways of life' and sources of identity;
- *environmental*...relating to our physical and biotic surroundings;
- *designed*...deliberately induced, not naturally evolving;
- *long-term*...relating to decades not years;
- *well-being*...not just material prosperity;
- *whole*...inclusive of all ages, both genders, all social groups; and
- *community*...here meaning people living or working in the relevant area.

Many of those terms are defined more rigorously later. But for now the above shorthand expressions will suffice to reveal the multi-faceted nature of rural development as it is currently and generally understood.

## LOCAL DEVELOPMENT

But why should 'rural development' be pursued principally at the local level? Why do rural programmes and plans and the projects that they contain need to relate not just to 'rural areas in general' but to this or that specific area? Why should machinery be put in place at the local level for determining and implementing rural development policies, programmes and projects? In short, why and how far should there be both 'decentralisation' (a shift of decision-making to 'lower levels') and 'territorialisation' (shift of focus from *sectors* such as education, transport and manufacturing to *areas*)? Setting aside for the present what 'local' might mean in terms of population size and geographical extent, there seem to be five main (and often overlapping) elements of the argument for specifically *local* development. (Useful references on this key issue include O'Cinneide and Cuddy (1992) and National Economic and Social Council (1994), both relating to rural Ireland, Buller (2000), on rural France, and, more generally, LEADER Observatory (1999a & b and 2001). The last-mentioned source posits the emergence of a distinctive 'European rural model' of development centred on the 'local area perspective'.

1. The first argument for *local* rural development relates to *local diversity*. Rural areas across Europe have much in common but they are far from being identical. Some have economies still dominated by agriculture; for others tourism, mineral extraction, retirement migration or manufacturing industry may be their principal vocation. Some may still be experiencing de-population, while for others it is rapid population growth and related social upheavals that characterise them. Some suffer from being 'too close' to metropolitan areas; for others it is remoteness that underlies their situation. Some are well-endowed with natural resources, others are not. So while all rural areas have, by definition, a scattered population and a landscape dominated by open countryside, their economic and social circumstances, their problems, needs and development potential will all vary greatly. It follows that the programmes that address their problems must be locally sensitive.
2. Second, *rural problems are interlocking*, and, in consequence, so must be both the measures to address them and the agencies involved. And the most effective way of achieving this may well be at an intermediate level, somewhere between the nation or region on the one hand, and the village or parish/commune on the other. It is at this level, the argument runs, that partnerships are best forged and co-ordination achieved or, to put it another way, that top-down priorities relating to sectors (such as healthcare, energy or specific industrial sectors) and bottom-up needs (across relatively homogeneous geographical areas) are best reconciled. As one Irish commentator put it, 'area-based partnerships have the potential to be the "central cog" that connects local needs and priorities with the "sectoral cogs" (sectoral programmes, funding and related agencies) which can supply the energy necessary for balanced and sustainable rural development' (Mannion, 1996: 12).
3. The third argument relates to *local identification and mobilisation*. It accepts that local people – both as individuals and collectively in groups, organisations and firms – are key resources in rural development, as sources of information, ideas, energy and enterprise. Such people will, however, only be enthused to participate if they feel that the venture at issue is clearly relevant to their concerns and that any contribution they make is likely to produce beneficial change. The more the area of operation is confined geographically and the more it is in some sense coherent rather than a hotch-potch of localities that happen to be in reasonable proximity to one another, the more this crucial resource of unpaid local energy is likely

to be forthcoming and sustained. So this argument is about building and mobilising social capital and drawing upon local knowledge and experience.

4. Fourth, there has been a growing sense that *adding value to local resources* is likely to provide a more secure and sustainable future for economic development than is a strategy involving excessive reliance upon imported materials and capital (even if, ironically, releasing that local added value often requires initial injections of non-local, for example EU, capital). This implies a need for a greater and more respectful understanding of local resources, in the broadest sense, and of their potential for creating new business opportunities. A second strand to this argument concerns the value of encouraging local purchasing by local people and organisations – a phenomenon graphically known as 'plugging the leaky bucket', with the implication that the local economic multiplier will be enhanced if money is recycled within the 'bucket' or local economy. Thus the argument is that local development driven by local people and institutions is more likely to foster both the adding of value to local resources and local purchasing.
5. The fifth argument has only really been voiced in recent years. It involves constructing a *defence against globalisation*. Globalisation (Bryden, 1998; Norberge-Hodge, 1999) entails the increased opening up of local economies to the cold blast of world competition. It arises particularly from the development and worldwide adoption of modern information and communication technologies, the global liberalisation of international trade and capital movements, the associated enhanced ability of multinational corporations to assemble capital wherever the costs of production are lowest and social and environmental restrictions are weakest, and international agreements that limit the power of national governments to directly bolster and protect the economies of their lagging areas. Thus cheese producers in Normandy, say, or cherry producers in Spain (see Case Study 7) have, increasingly, to accept that very similar produce from places thousands of miles away is occupying 'their' shelf space in their nation's supermarkets. One response to this has been to deliberately accentuate and proclaim local diversity, to foster in each local area a distinctiveness and, thereby, a 'niche' at least in the mind of the consumer. The urgency of developing and marketing local identity and distinctive quality products and services linked to it is, then, another case for rural development being pursued at the local level, and it is one of growing importance – as recently argued in Ray's consideration of what he terms 'culture economies' (Ray, 2001).

For some or all of these reasons the 1990s and the early part of the twenty-first century have witnessed a proliferation of local development initiatives in both urban and rural areas. These have included, in England, various 'Rural Development' and 'Single Regeneration Budget' programmes (Cherrett, 1999) and the recently launched local authority 'community strategies', in Ireland local Programmes for Economic and Social Progress (Keane, 1998), in France the 'contrats de pays' and 'intercommunal syndicates' (Buller, 2000), in Finland the 'POMO' or 'Programmes of Rural Development Based on Local Initiative' (Kahila, 1999) to take just a few nation-specific examples. And at the pan-European level we have, for example, LEADER (see Case Study 1) and the 'Territorial Employment Pacts' (European Commission, 1997).

Given those arguments and that experience, we may now define *local development* (whether urban or rural) as 'the pursuit of development – as previously defined – at a local scale with the aim of addressing local concerns, adding value to local resources – whether material, human or symbolic – and mobilising local actors – whether people, groups or agencies'. It follows that *local rural development* – the core focus of this book – is local development as nuanced by rurality.

Our definition of local development is very much in keeping with the remarks of an Irish commentator who observed some years ago that it is 'more than a scaling down of interventions previously organised from the top by centralised policy making units ... it is a radical response that seeks to achieve new objectives in relation to the development process by focusing on such concepts as multi-dimensionality, integration, co-ordination, subsidiarity and sustainability' (Walsh, 1995: 1). In that regard Walsh suggested three specific tasks for local development, namely:

- overcoming 'market failures' (meaning doing socially useful things that are generally unattractive to the market, such as delivering services to a scattered population and integrating environmental conservation and economic development programmes);
- improving 'local capacity' (meaning the ability and readiness of people and organisations to engage in development initiatives); and
- facilitating 'local empowerment' (meaning giving local 'actors' more power to influence what happens in their locality).

In similar vein, another influential Irish critique of 'rural development' stressed the importance of its pursuit at the local level where each of the following might be most effectively achieved (National Economic and Social Council, 1994: xiii–xiv):

- 'pre-development' – meaning capacity building and the animation of local groups;
- the operation of area-based partnerships;
- the adoption of a strategic planning approach;
- the fostering of innovative projects and methods;
- the reduction of social exclusion;
- the development of enterprise; and
- the promotion of community and group projects.

But how local is 'local'? Obviously the answer to that question must depend to a considerable degree on such local features as the population density, resource base and administrative structure, but some guidance is possible (see, for example, National Economic and Social Council, 1994 part 2, chapter 13, and various LEADER documentation). The main point is that to best pursue the sorts of objectives outlined above, the local area should be small enough to sustain a 'sense of place', the willingness of local people to get involved and the prospect of a real integration of individual initiatives, but also large enough to afford certain economies of scale in management and service delivery and the likely availability locally of a sufficient range and quality of expertise. And larger areas are also likely to embrace at least one small or medium-sized town which will bring its own benefits to the development process. Having weighed such factors, Ireland's National Economic and Social Council (1994) suggested the 15 to 25,000 population range as being generally preferable, while, as explained in the case study which follows, the experience of LEADER indicates an upper limit of 100,000 and an average of around 50,000. But as important as size, if not more so, is the desirability of focusing on reasonably coherent areas enjoying some measure of shared identity. That, and a firm preference for areas bigger than the individual English parish or French rural commune, and smaller than the English county or French *département*, is about as far as commentators on this subject generally go.

Interestingly, as this advice on local, rural development was being crystallised by and for practitioners, Marsden (1999) produced a quite similar list of research priorities for rural social scientists. He stressed a need for a greater understanding of diversity within and between rural areas; of the ways of achieving area-based, holistic and integrated rural development; of the emerging new forms of local governance and partnership; of citizenship, capacity building and the mobilisation of local populations; and of the capacity of rural areas for sustainable endogenous development.

## THE PRESENT BOOK

So much for the arguments for 'rural development' in advanced western economies and for its pursuit at the local level. The chapters that follow take many of the concepts and ideas considered above and seek to explain them and their role in the development process. Collectively, they provide some 'building blocks' for a theory of 'local rural development' though they do not in themselves comprise such a theory.

What such theorisation would involve ('theory' being in essence an attempt to explain and predict real-world phenomena in as concise and elegant a form as possible) is a rigorous formulation of how these and other 'building blocks' inter-relate in practice and of how and how successfully they contribute to 'local rural development' on the ground. This would require, among other things, the careful analysis of hundreds of actual exercises in both the planning and delivery of 'local rural development' and the teasing out of the relative contribution made to them by these and other elements whether individually or in tandem. To do that would be an ambitious, but not impossible, undertaking. All we can do in this book is to offer some material for that theorisation and some clarification of its significance. Also offered, however, is a good deal of practical advice to development agencies and local groups wanting to know 'how do we set about this?'

The 13 chapters that follow each focus on one particular concept or issue in local rural development – many of them already introduced earlier in the present chapter. Nine relate to broad principles or goals; four to the key steps to be taken at the area level to address or achieve them. But the distinction is not clear-cut for the simple reason that in local development the 'process is also the product' and that the 'product adds fuel to the process'. For example, several goals of the development process, such as 'sustainability', 'community' and 'social inclusion' (the foci of Chapters 2, 6 and 7 respectively), *themselves* provide a spur to further development. And necessary management tools like 'strategic planning' (Chapter 12) and 'evaluation' (Chapter 14) *themselves* help to develop the people and organisations involved in them – or at least they should do. This is nothing more than a reassuring confirmation of the author's conviction that, properly conceived and undertaken, development is a 'virtuous spiral' in which everything positively affects everything else. But in another sense it confirms a need to resist premature assertions about how the whole edifice hangs together. Thus the reader is invited to treat what follows as a series of individual essays, each devoted to a core theme of local rural development, and to ponder for him/herself the variety of ways in which they might be hooked together both intellectually and in the shaping of practice.

The structure *within* each chapter is common: first, an explanation of the meaning of the concept and of its significance in local rural development; second, a brief reflection on some key issues surrounding it; third, an outline of the 'toolkit' available to help attain or undertake it; and fourth, two case studies to give more real-world substance to a subject which might otherwise appear too abstract.

**Chapter 2** is devoted to an overarching, perhaps *the* overarching, concept or guiding principle of local rural development namely *sustainability*. This is defined to embrace the conservation of an area's 'manufactured, human and social capital' as well as that which is inherently natural or 'environmental'. The chapter suggests various elements of 'good practice' in the pursuit of sustainability within local rural development.

**Chapter 3** stresses the importance of *innovation* – of doing something different, of 'breaking the mould' – in the local rural development process, and explores the circumstances in which innovation is most readily adopted by the relevant 'actors'.

**Chapter 4** is concerned with the *adding of value* to local resources as a strategy for local rural development. It links with the earlier chapter on 'sustainability' and with a later one on 'diagnosis', all three focusing on local resources but from different perspectives.

**Chapter 5** focuses on business and community *entrepreneurs* – those crucial people with an eye for opportunity, a desire to achieve and a readiness to take risks. Economic and community development cannot take place without them and their careful cultivation is an essential element of strategies to foster local development.

**Chapter 6** considers *community*, something which is difficult to define but clearly valued by residents and service delivery agencies alike. Often weak if not actually lacking in specific localities, and generally under threat, it is also central to the process of local development.

**Chapter 7** is devoted to *social inclusion*. Its focus is on the mechanisms which exclude many rural people from the lifestyles of the majority and on the ways in which local rural development can best address them.

**Chapter 8** explores *accessibility*, meaning the ability of people to reach the things that are important to them. It argues that there is much that local development strategies can do to improve the accessibility enjoyed by disadvantaged local people, by influencing transport, communications and service delivery, though the challenge is to do so by galvanising development and not simply by 'filling gaps'.

The next two chapters consider the local machinery and 'human and social capital' needed to devise and carry out local rural development programmes. Thus **Chapter 9** is devoted to *partnerships*, the formal structures needed if the relevant actors from the statutory, private, voluntary and community sectors are to be harnessed together to work with common purpose. And **Chapter 10** explores the 'why, how, how far and who?' of *community involvement* in local development.

The last four substantive chapters all relate to the business of bringing about local development in an efficient and effective way. They suggest a sequential process. Thus **Chapter 11** is concerned with *diagnosis*, or the task of establishing the baseline conditions of an area prior to shaping a development programme for it – that 'baseline' embracing its resources, opportunities, problems, needs and constraints. **Chapter 12** is devoted to *strategic planning*, the process whereby the actors in a local area collectively build on that diagnosis to define a vision, set objectives and devise a coherent set of associated measures to resolve the problems identified. **Chapter 13** argues that no amount of elegant planning can promote development if corresponding attention is not paid to the *implementation* of plans on the ground over a sustained period of time, and it focuses on the types of intervention that are possible and on how they can best be put into effect. Finally, **Chapter 14** deals with *evaluation*, whereby the achievements of a programme or its constituent projects are periodically assessed and explanations sought for any deviations from the intended plan.

What these four 'technical' chapters (Chapters 11 to 14) share is a conviction that technical expertise is not enough. Delegating the four tasks to 'experts' standing aside from the messy political process of making development happen on the ground is both to weaken the tasks' effectiveness and to miss a golden opportunity. Their effectiveness depends as much on harnessing the talents and wisdom of local people and groups as it does on the experts' technical competence in gathering, analysing and reporting information. And the 'golden opportunity' so easily squandered is the chance that they offer to *develop* those local people and groups through their being intimately involved in real and important exercises that promise to excite and stimulate them. Again 'the process is part of the product' or, more precisely, the process can and should of itself yield relevant products.

Thus the whole book is about the promotion of local rural development – what? why? and how?

## CASE STUDY 1: EUROPE'S LEADER PROGRAMME

The challenge of trying to put into practice some of these principles of 'local rural development' is well illustrated with reference to the European Unions' LEADER Programme (*Liaisons Entre Actions de Développement de l'Economie Rurale*). That programme was born of Brussels' growing realisation (Commission of the European Communities, 1988) that in the 1990s it should more fully respect the diversity of rural Europe, complement narrow agricultural policies with others more comprehensively 'rural' in their scope, and give *local* actors and agencies more responsibility for devising and managing them than the national agencies with which the Commission had normally worked hitherto.

The LEADER Programme was launched initially for three years (1992–94). Then, having proved its worth, it was rolled forward with some relatively minor changes but on a larger scale as LEADER II (1995–2001) and again as 'LEADER Plus' (2002–6).[1] At the time of writing the details of the LEADER Plus programme remain somewhat sketchy but LEADER I involved 217 local areas within Europe's designated disadvantaged regions receiving funding to devise and implement local development programmes, and in LEADER II this increased to some 900 areas, ranging from Italy's 185, via the UK's 66 to Luxembourg's two.

Many of these LEADER II areas were carried forward from LEADER I with the result that about 200 local rural areas, spread across the 15 member states, had by 2001 had some nine or ten years' experience of practising local rural development, and so some interim assessment is possible. (The growing literature on this subject includes LEADER Observatory, 1999b and c; Ray, 1996a and b, 1998, 2000a, 2001; and a special issue of *Sociologia Ruralis* (2000).)

But it should first be noted that these local development programmes have not been lavishly funded; the European Union's LEADER II allocation of little more than £1 billion spread across about 900 areas and over six years was only about 2 per cent of all its 'Structural Funds' expenditure in its priority geographical areas – most of it going directly into agricultural support. Thus for LEADER II there was an average yearly allocation to the 'local action groups' of only some £200,000, though it was a requirement that this funding be matched by roughly equivalent money from national and local sources. Thus the hope was that local development would be triggered through a judicious programme of backing well-chosen small projects and by the innovative process of local action to which we will now turn.

There have been seven key elements of 'the LEADER approach' to local rural development (LEADER Observatory 1999a and b), two of which, relating to transnational collaboration between local LEADER groups and to the EU/nation/locality financial management procedures, will not concern us here. The other five, however, were and are fundamental:

1. The *area-based approach*, or the 'territorialisation of development initiatives', was substantially new to rural Europe when LEADER was launched in 1992, except in France (Buller, 2000) and in Britain and Ireland, where the national governments had initiated such an approach in the 1980s (Westholm, Moseley and Stenlas, 1999). Its rationale in the LEADER programme reflected points made at the start of this chapter – notably the championing of diversity, a determination to mobilise local people and organisations and the need to address inter-related problems in a way impossible at the national or even regional levels. The European Commission specified that the LEADER areas should have some real local identity, rather than simply respect established administrative boundaries, and should not normally have more than 100,000 inhabitants. In the case of the 20 English LEADER II areas, for example, their average geographical extent was some 1000 sq. km and their average population 52,000, and all but three crossed district and sometimes county boundaries (Ray, 1998).
2. The *bottom-up approach* has involved placing a high premium on the active participation of people living within the selected areas. This has meant partly the formation of *ad hoc* 'local action groups' to manage the programmes (see below), partly a requirement that local people be consulted and involved in shaping the development programme for the area and partly an expectation that most of the project proposals vying for support would come 'up' from people, businesses and organisations at the most local level rather than 'down' from central or local

government. Ray (2000a) has termed this an 'anarchic post-modern approach to intervention', and certainly it seems to have proved rather hard to palate for some regional and national governments across the continent. But allegations that such an approach is fundamentally 'undemocratic' in showing scant respect for the traditional organs of 'representative democracy' are commonly countered with arguments that, in fact, it fosters a richer 'participative democracy' (Ray, 1998).

3. The *local partnership approach* has involved the creation or consolidation of *local action groups* to devise and manage the local LEADER programmes, drawing up 'local action plans' to bed their work in local needs and resources and determining how the limited funds available should be disbursed between competing project applicants. In Britain, at least, these local action groups have generally been widely drawn from local business, the local authorities and voluntary and community organisations, and have been serviced by a salaried 'project co-ordinator' and one or more field staff. This 'partnership approach' has certainly worked, sometimes exceedingly well, but there have been frequent criticisms that some local action groups have focused excessively on project selection at the expense of championing the 'bigger picture' of integrated local development, or else have been effectively 'in the pocket' of state agencies or the local authorities. There is also clear evidence (see for example Case Studies 25 and 27) that many have grown frustrated at the complexity of the procedures for drawing down funds for even very modest projects, and more generally at the way over-cautious regional or national bodies have resisted 'letting go'.
4. *An emphasis on innovation.* 'Innovation' is a word that recurs time and again in the official LEADER literature – not surprisingly, as a major objective of LEADER has been to test out new ways of addressing rural problems in the hope that some would subsequently be 'mainstreamed' (meaning incorporated into European or national 'mainstream programmes and policies'). Thus the Commission has insisted on an innovative management approach at local level (see the previous paragraph) and also on innovative projects getting the bulk of the project funding. The Brussels-based 'LEADER Observatory', charged with helping the 900 or so local action groups of LEADER II to 'network' and thereby exchange good practice, laid great weight on monitoring and disseminating 'innovative practice' (see for example LEADER Observatory, 1999d, some of this work being summarised in Moseley, 2000a). But while substantial innovation has undoubtedly occurred – for example a host of locally novel ways of adding value to local agricultural produce or of exploiting an area's cultural heritage – it is also clear that 'more of the same' has been equally apparent. Kearney *et al.* (1994) noted, for example, in their evaluation of Ireland's LEADER I programme that a disproportionate amount of funding went into yet more 'run of the mill' bed and breakfast accommodation rather than into something really different or special.

5. *An emphasis on integration.* The final hallmark of the 'LEADER approach' has been a wish that the local programmes be not just multi-sectoral, relating, for example, to vocational training, rural tourism and the promotion and marketing of the local area, but genuinely *integrated*. An example of the latter would be training courses provided for farmers who are keen to diversify, linked to grants to help create on-farm accommodation and linked also to the marketing of the area as a destination for rural tourism. Hard data on this is hard to come by, but the author's impression is that, though some striking successes have occurred, such integration has generally proved to be elusive or else overlooked by local action groups anxious to treat individual applications for project funding on their merits.

In conclusion, and as befits a programme designed to champion local diversity, the character and success of LEADER has varied considerably across the 15 member states of the European Union. (For some national reviews in the English language see the special issue of *Sociologia Ruralis* (2000) devoted to LEADER. For brief reviews in this book of specific local LEADER programmes, see Case Studies 2, 7, 18, 25 and 27.) But there is some welcome evidence that LEADER has indeed, as initially hoped, served as a 'model programme', with its essential features, as listed above, increasingly replicated in other rural development programmes. Examples of replica programmes are PRODER in Spain and the POMO in Finland (Westholm *et al.*, 1999).

Whether, and how far, LEADER has genuinely spurred integrated, sustainable local development, rather than merely supported a worthwhile collection of small, one-off projects, is, however, a much harder question to answer. The answer is probably: 'to some extent and to an extent that varies greatly from place to place'. The cross-national PRIDE enquiry into the impact of local rural partnerships, discussed in Chapter 9 and reported more fully in Moseley (2001 and 2003), provided some encouraging evidence in that respect, but even in 1999–2000, the time of that research, the indications were that it was too soon to say with real certainty just how influential LEADER had been. Local rural development is a long and often nebulous process.

---

### CASE STUDY 2: SPARC – THE SOUTH PEMBROKESHIRE PARTNERSHIP FOR ACTION WITH RURAL COMMUNITIES

---

SPARC[2] was created in 1991 with a mission to involve local people in improving the social and economic well-being and enhancing the environment of rural South Pembrokeshire, an area of Wales with some 43,000 people living mainly in 35 small towns and villages and isolated farmsteads spread across an area of about 400 sq. km. It is an area with, historically, a strong reliance on agriculture, forestry and seasonal, almost entirely coastal, tourism and with serious problems of low incomes, high unemployment and youth out-migration. (Midmore *et al.*, 1994; UK LEADER II Network, 2000).

SPARC was a local development partnership with a council of management elected by an assembly or 'consultative committee' which, in turn, drew its membership from four 'constituencies'. These were the area's local authorities, SPARC's funding agencies, various advisory bodies and, most interestingly, a network of local village-level community associations. The latter, which together covered virtually the whole area and to which all residents were entitled to belong, brought a genuine element of 'grass roots democracy' to the development partnership which served them.

As for staff, SPARC employed six development workers each with a sectoral remit – farm support, countryside, community tourism, business development and training, local food development and community support – plus an administrative officer and a co-ordinator who led the whole team and was responsible to the council of management. Funding came partly from the LEADER I and LEADER II programmes – which, together with LEADER Plus, will, by 2006, have supported many of the local development initiatives of SPARC and its successor body over a period of 14 years – and partly from a range of other EU, national and local sources.

The village-based community associations were central to SPARC's underlying aim of giving local people the chance to develop their own communities economically, socially, environmentally and culturally. Each association undertook at least one 'community appraisal' – a questionnaire- based survey of local people's needs and wants designed to provide the basis of a Village Action Plan which would set out priorities for the development of that community. In the larger villages and small towns much of the preparation of these action plans was undertaken by thematic working groups, focusing on employment, local services, the environment and other key issues emerging in the appraisal and translating the appraisal's findings into practical initiatives to be endorsed or modified at periodic conferences of local residents.

Once the 'action plan' had been adopted locally, SPARC then worked to support the relevant community in taking it forward, helping them to network with the agencies that could assist them, providing training and practical help of various kinds and part-funding new businesses and community projects. SPARC also developed and, in large measure, implemented a number of area-wide plans which addressed more strategically the needs identified in the various village action plans.

Before looking at SPARC's work on the ground it should be noted that the very process of conducting appraisals and developing action plans in an inclusive, democratic fashion produced its own developmental spin-off. It built up the awareness, confidence and skills of local people and an appreciation of the benefits of working in partnership with neighbouring communities and outside agencies. It also served to persuade those agencies of the benefits to them of collaborating with local people.

What happened on the ground as a result of all this? Five particular projects, all pursued, in part, with EU funding, are indicative:

- *Makers of Wales* is a national campaign to celebrate Wales' heritage which SPARC turned to the advantage of local communities by helping them get funding for a number of conservation and cultural tourism initiatives including the restoration of historic sites, the upgrading and way-marking of footpaths and the production of promotional literature, interpretation panels etc.
- The *Local Products Initiative* recognised the economic benefits of purchasing from local sources. Networks linking local food producers and purchasers were created with an emphasis on encouraging the local tourism industry to buy locally. Training programmes were provided for local women wishing to learn new ways of adding value to local resources.
- The *Quality in Business* initiative served existing and potential small businesses by providing locally based advice and training, incubator premises, environmental and energy audits and small grants to enhance business efficiency.
- *Supporting Communities in South Pembrokeshire* encouraged a variety of village-based activities, such as the more efficient use of village halls, the provision of childcare facilities, local networking using information technology, community enterprises and conservation schemes identified by local people.
- The *Demonstration Farm Review and Development Scheme* encouraged the development of whole-farm business plans linking training, diversification and conservation audits to funding for business and environmental improvements.

---

Recalling the various tenets of 'local rural development' discussed earlier in the chapter – notably its area-based focus, the adding of value to local products, the promotion of community involvement, local partnership, innovation and integration – it is not difficult to see why SPARC gained a reputation across Europe as a commendable model of the LEADER approach. Indeed, for that same reason it provides a good entrée into the rest of this book which will examine more carefully those various precepts.

### NOTES

1. These dates reflect action on the ground and not the somewhat fictional timetable set out in EU pronouncements.
2. 'SPARC' was formally wound up in 2001, with PLANED emerging from it to undertake similar work but over a wider area of Pembrokeshire.

# Part I

# Principles

# 2

# Sustainability: Respecting the Long Term

For one sweet grape who will the vine destroy?

Shakespeare, *The Rape of Lucretia*

## DEFINITIONS AND IMPORTANCE

Today, no regional or rural development strategy fails to pay at least some attention to the goal of 'sustainability'. Some embrace the concept with real enthusiasm and allow it to inform the whole conception of what the strategy is about as well as the selection and fashioning of programmes and projects for implementation. Others are more grudging, construing 'sustainability' narrowly as the protection or conservation of the physical and biotic environment. In the latter conception, environmental protection is treated as a counterbalance to the main business of developing the economy, creating jobs and providing local people with the services and facilities that they want. It is treated as a 'luxury good' which the development agencies are prepared to buy into – but to an extent limited by the key economic objectives. (For discussions of the reconciliation of 'rural development' and 'sustainability' see, for example, Lowe and Murdoch, 1993; Bryden, 1994; Bryden *et al.*, 1998; Macdonald *et al.*, 1998; and LEADER Observatory, 2000a.)

In fact such an interpretation completely misses two fundamental points: first, that 'the environment' is just one of several resources that demand to be conserved and, if possible, enhanced for future generations; and second, that conservation and, development are not two opposed objectives that need to be balanced, but rather, if properly addressed, they are complementary and mutually reinforcing. To bring them into a state of mutual reinforcement, rather than mere 'balance' is really what sustainable development is about.

Two definitions are necessary: sustainability and sustainable development. Based on some work by the Cheltenham Observatory (1998), which is reviewed at greater length below, we suggest that *sustainability* be defined as 'the capacity for continuance into the long-term future'. It is built around the notion of 'conserving capital', but in a particular sense which, following earlier work by Ekins and his colleagues, sees that capital as coming in four forms[1]:

- *environmental capital*, which comprises stocks and flows of energy and matter, and the physical states, such as climatic conditions or ecosystems, to which they give rise;
- *human capital*, which comprises the ability of individual people to do productive work, whether paid or unpaid, and therefore includes their physical and mental health, their strength and stamina, their knowledge, skills, motivation and attitudes;
- *social capital*, which relates not to *individual* people but to the social structures, institutions and shared values which enable individuals to maintain and develop their human capital and to be productive. It therefore embraces firms, trade unions, families, communities, informal friendship networks, voluntary organisations, legal and political systems, educational institutions, the health service, financial institutions, systems of property rights etc.; and
- *manufactured capital*, which comprises material goods such as tools, machines, buildings and infrastructure, all of which contribute to the production process without becoming embodied in its output.

Some 'capital' may be hard to categorise. For example, is 'landscape' environmental or manufactured capital? In Europe it is generally a mixture of the two. But the point is that 'capital can be conveyed as bequests between generations' (Selman, 1996: 12, citing earlier work by Pearce), and that the destruction of capital by inappropriate development and without adequate replacement is to create a state of affairs that cannot be sustained indefinitely. Particularly serious is the destruction of non-renewable capital, most of it in the 'environmental' category, but some can be cultural and therefore man-made, such as a minority language. Notwithstanding that, however, the underlying argument of sustainable development is that all four capitals should be respected. Thus, in the context of local development, we should be as concerned with rural housing and planning policies which effectively destroy locally based extended families by pricing young people out of the housing market (thereby destroying some valuable 'social capital') as much as with a proposed industrial estate that would destroy a valued, but not unique, wildlife habitat.

If 'sustainability' is the destination, *sustainable development* is the journey. We may define it using the celebrated form of words coined by the Brundtland Commission in 1987: 'development that meets the needs of the present without compromising the ability of future generations to meet their own needs', or the expression preferred by the Cheltenham Observatory (1998: n.p.:) a dynamic process that enables all people to realise their potential and to improve their quality of life in ways that simultaneously protect and enhance the Earth's life support systems', the last four words of the latter definition being tougher than Brundtland on the imperative of respecting biophysical limits.

Valuably, those definitions each make clear that sustainable development is not only to do with the well-being of our children and grandchildren; it is also concerned with the problems of here and now. To put it differently, they express a need to strive for *intra*-generational equity (a fairer today) as well as *inter*-generational equity (a fairer tomorrow). Given that, it seems reasonable to accept the 'explicit adoption of sustainability as a touchstone of urban and rural development' (a phrase coined Selman, 1996: 1). In other words, to see 'sustainability as 'a', if not 'the', unifying concept underlying the whole of this book. Thus every subsequent chapter of this book is in effect exploring, and indeed advocating, sustainable rural development, albeit from different perspectives.

## SOME KEY ISSUES

The sustainability literature, however, also advances another 'fundamental principle', in addition to those of intra- and inter-generational equity. It is that of 'transfrontier responsibility' – the notion that sustainability 'here' cannot be achieved at the expense of eroding sustainability 'there'. This is important in the context of local development, the focus of this book, since it raises questions about policies that may seem 'green' – for example the development of eco-tourism based on the sustainable use of local environmental and manufactured capital, only for that tourism to rely on tourists making extravagant demands on fossil fuels by travelling hundreds or thousands of miles for the experience. There is a dilemma of local development here which is far from being resolved satisfactorily.

Turning now explicitly to the appropriateness of local development as a mechanism of *sustainable* development, we agree with Selman (1996: 3) that 'the local arena is often the crucial arena in which sustainability may be pursued'. Some see that arena as but a side-show to the global struggle, but that level of pessimism seems too negative for several reasons.

First, sustainability requires that social, economic and environmental issues be considered together rather than in separate boxes – and the local scale often offers greater scope for so doing than does the regional or national. Indeed, local development is inherently about holistic planning and action. Second, it is often at the local level that conflicts between competing objectives are best resolved because locally generated solutions are usually the product of genuine face-to-face debate between the parties most likely to be affected (a justification of the so-called 'subsidiarity' principle of transferring as much power as possible down the decision-making hierarchy). Third, local areas can, depending of course on their scale, coherence, culture etc., yield crucial local knowledge as well as a climate of greater mutual trust and shared responsibility. Fourth, local development can provide a convenient vehicle for concerned people to 'do something', rather than be mere spectators at the 'sustainability circus'. Fifth, any actual or potential polluters or extravagant users of common resources will be more likely to be deterred by the impending opprobrium of their neighbours and peers than by the disapproval of bureaucrats located 200 miles away; by the same token those disbursing development funds will also be locally accountable. Sixth, and perhaps most important, the very process of undertaking local development, and not simply the projects that it champions, can and indeed should advance sustainability, particularly by building up both human and social capital as a by-product of its work.

For all these reasons, local development, whether rural or urban, is well placed to embrace the principles of sustainable development – but this need not happen automatically. We turn, therefore, to a consideration of what 'local rural development' must do to promote sustainability; what is 'good practice'? Posing that question requires us to consider both the way that local rural development is interpreted and pursued, and the projects that it champions and supports. We will look at each in turn.

## GOOD PRACTICE IN THE PURSUIT OF SUSTAINABLE LOCAL DEVELOPMENT

Sustainable local development must be based on a vision relating more to long-term human welfare than to maximising the production of goods and services or the crude creation of as many jobs as possible. As suggested earlier, this implies internalising into the culture of its decision-making processes respect for the 'four capitals' and the need to protect and, if possible, increase them locally. Related to this is the need to jettison an often implicit model of seeking to strike a balance between conservation and development, and to replace it with one of setting in train a spiral of development and conservation, each supporting the other. In turn, this means seeking out 'win–win' initiatives which make use of one or more of the 'four capitals' in a way that enhances human welfare, while at the same time enhancing, or at least not depleting, them to an unacceptable degree.

Attaining that goal is significantly helped by making local development as integrated and inclusive as possible. 'Integrated' in the sense of embracing, simultaneously, economic, social, cultural and environmental issues and resources, and 'inclusive' in the sense of involving fully the various stakeholders likely to be affected by the development programme. And pursuing an integrated and inclusive development programme is likely not only to lead to the selection and support of a more sustainable portfolio of projects than might otherwise be the case, but also to increase the area's stock of both human and social capital. That stock is likely to be further augmented by explicitly championing a 'bottom-up' model of development, with many of the initiatives to be fostered coming up from local people rather than down from the experts.

Also important is the development and use of pragmatic but sound procedures and criteria for assessing the sustainability implications of proposed and actual projects. (Case Study 4 presents one such approach). Indeed, more generally, there is a need to develop a way of working which poses searching questions about sustainability (in effect, about the likely consequences for the area's inherited capital) at *all* the various stages of devising and executing the local development programme – namely diagnosis, plan-making, project selection, implementation, monitoring and evaluation.

## GOOD PRACTICE IN THE CHOICE OF SUSTAINABLE PROJECTS

Going on from procedures, what of the actual projects that will tend to figure in sustainable local development strategies? Both of the two case studies below present a number of examples of what are generally sustainable projects, but it is useful to try first to establish some general principles.

Always remembering the economist's common caveat, 'other things being equal' (in other words acknowledging that there may be special circumstances that on occasions invalidate the general principle), sustainable projects tend nevertheless to:

- emphasise the re-use or recycling of redundant 'manufactured capital' and land rather than its abandonment or replacement;
- reduce the use of personal motorised transport, indeed of motor vehicles generally;
- replace some unnecessary movement of people and commodities by telecommunication technology;
- promote the very local provision of employment and services so as to reduce the length of journeys to work, to shop etc.;
- reduce per capita energy consumption, especially that derived from fossil fuels;
- protect areas of critical landscape or wildlife value;
- incorporate participatory and inclusive decision-making;
- build up human and social capital;
- use local labour and materials rather than that brought in from afar;
- increase the self-reliance and self-sufficiency of local economies by reducing imports and exports to/from the rest of the world;
- develop 'win–win' situations in which local resources are used for economic or social development while being simultaneously enhanced; and
- not pose problems of waste disposal

That list is not exhaustive, nor indeed problem-free. By way of example, three possible contradictions may be admitted. The first concerns the very local (inevitably meaning 'scattered') provision of employment which may, as people exercise choice in the job market, lead perversely to longer, not shorter, journeys to work, and to more, not less, use of personal rather than public transport. The second concerns local tourism development strategies that commendably try to add value to, and simultaneously conserve, distinctive local assets but do so only at the price of persuading a widely dispersed clientele to make long, energy-consuming trips to come and enjoy what is on offer. The third example is a variant of the second. Increasingly, and commendably, local areas seek to produce and market quality products based on the sensitive exploitation of a local resource; for example Case Study 7 in Chapter 4, concerns the production of cherry brandy in a remote part of Spain for transport and sale throughout Europe. But as with the tourism example, this does raise questions about 'transfrontier responsibility' and the possible export of unsustainable practice to places outside the local area.

## THE TOOLKIT

Moving on to consider specific ways in which local development agencies and other local actors can actually promote sustainability in their plans and actions, it is important first to note the danger of what might be called 'project-ism' – the assumption that promoting discrete one-off projects comprises all of what is possible or necessary to achieve one's goals, a point reinforced in the 'implementation' chapter below. In fact, it is often just as important to seek to influence external agencies and 'mainstream programmes' that have a significant impact on the area in question. This is one of the reasons why rural development partnerships, incorporating a variety of regional and national, as well as truly 'local', actors, are important. The latter may need to seek to persuade the former – the transport providers, the land-use planners, the agricultural agencies, the housing providers – to respect 'sustainability' in their day-to-day dealings with the local area in question. Otherwise 'sustainable projects' may prove to be somewhat insignificant additions to a fundamentally unsustainable situation.

Going on to consider specific tools which local development agencies might deploy to promote sustainability, several have already been implied above. So here we will simply set out in a slightly amended form (Table 2.1) the list of 'sustainability criteria' defined by one research project concerned to promote good practice in rural development (Cheltenham Observatory, 1998). Here we modify those criteria slightly to express them as 'goals', and list alongside them some examples of local development practice that relate favourably to them. Note that the list embraces all four 'capitals' as previously defined; indeed only goals 1, 2 and 3 explicitly relate to the physical and biotic environment.

**TABLE 2.1** ***Some goals and tools for local sustainable development (drawing upon Cheltenham Observatory, 1998 and other sources, and placing the goals and tools in no particular order)***

| A goal for sustainable local rural development | Some examples of tools and projects that may be applied in rural areas |
|---|---|
| 1. To avoid the **depletion of non-renewable resources** at a rate faster than allows for the development of appropriate substitutes | Emphasis on re-use, repair, recycling and reconditioning. Also the development of substitutes for some resources, or the use of more efficient technologies.<br>Examples:<br>transport – projects and programmes that reduce the use of the motor car; grants for the re-use of redundant buildings such as former defence establishments;<br>development of woodland resources for biomass;<br>avoidance of industrial projects which make heavy use of fossil fuels, minerals or aggregates. |
| 2. To avoid **emissions into air, soil and water** that exceed the capacity of natural systems to absorb or neutralise harmful effects | Routine application of project appraisal systems that screen for $CO_2$ emissions, discharges into water courses, etc.<br>Examples:<br>application of cleaner technologies in farming practice;<br>promotion of local recycling schemes. |
| 3. To maintain, enhance or restore the diversity and productivity of natural **ecosystems** | Support for projects which conserve ecosystems and wildlife habitats, either directly or in conjunction with environmentally friendly recreation or tourism projects.<br>Examples:<br>projects to re-establish native species at risk;<br>projects to develop tourism linked to the protection or restoration of semi-natural habitats. |
| 4. To develop the skills, awareness, knowledge, health and motivation of **local people** so that they have greater access to productive and satisfying work and participate in other developmental activities | Various labour market and non-labour market initiatives that seek to develop the human resource.<br>Examples:<br>training programmes in information and communication technology for currently 'excluded' people;<br>capacity building in the voluntary and community sectors, use of schools for wider community education; |

*(Continued)*

**TABLE 2.1** ***Continued***

| A goal for sustainable local rural development | Some examples of tools and projects that may be applied in rural areas |
|---|---|
| | involvement of a wide range of people in participatory processes. |
| 5. To conserve physical infrastructure and other **manufactured capital** | An emphasis on restoring and conserving redundant buildings, machinery, landscape features, transport routeways, etc. Examples: grants for the refurbishment and re-use of redundant agricultural buildings; bringing empty or under-occupied dwellings into use to reduce the need for new housing on greenfield sites. |
| 6. To maintain or develop **technologies** which ensure production and consumption that is resource efficient | Promoting respect for consequences of competing technologies which go beyond the 'bottom line'. Examples: promoting the energy efficient use of buildings and energy efficient transport; preserving land drainage systems that sustain diverse habitats and ecosystems; generally favouring more labour-intensive, less capital-intensive agriculture. |
| 7. To develop effective and equitable **organisations and institutions and management systems** that contribute to economic development which respects the conservation of natural, man-made and human resources | Promoting, in the private, public, voluntary and community sectors, structures and systems of management more likely use resources in a sustainable way. Examples: creation or enhancement of mutual societies, co-operatives, community-run enterprises; creating and sustaining local development partnerships and networks. |
| 8. To maintain and improve formal and informal **systems of justice, democracy and governance** that promote social cohesion and justice | Promoting local socio-political developments that build up an area's social capital. Examples: increasing participation in the local electoral process; trying new ways of enhancing genuine popular involvement in local decision-making (e.g. 'planning for real' exercises, community health councils, rural transport forums); |

*(Continued)*

**TABLE 2.1** ***Continued***

| A goal for sustainable local rural development | Some examples of tools and projects that may be applied in rural areas |
|---|---|
| | support for socially inclusive community-based groups. |
| 9. To maintain and improve **homes, families, neighbourhoods and communities** that provide safe and convivial environments | Recognising that just and mutually respectful human relationships, community identity and participation all contribute to sustainability<br>Examples:<br>community involvement in crime prevention;<br>policies for affordable housing that keep three-generation families together in the same community;<br>local environmental action projects that also generate community solidarity. |
| 10. To maintain and encourage diverse **cultures, traditions and activities** which enhance continuity with the past, the celebration of 'local place' and local pride | Celebrating 'local place' and heritage and fostering a sense of local pride.<br>Examples:<br>projects concerned with local cuisine, arts and crafts, music and dance, festivals, cultural landscapes (i.e. distinctive landscapes 'built by hand') and townscapes. |

Many of the above examples of tools and of good practice are developed in greater detail in Case Study 3, and in other chapters.

### *THE 'LEAKY BUCKET' ANALOGY*

The above attempt to sketch out a toolkit for the local promotion of sustainability sits comfortably with a conceptualisation of local economies which equates them with 'leaky buckets' (e.g. Douglas, 1994; New Economics Foundation, 2001).

The essential idea is that when, say, £1000 enters a local economy (or flows into the 'bucket') – for example payment for some goods that have been exported, or a pension payment to someone who has retired to the area, or the salary cheque of a local resident who commutes to work outside the area, or expenditure by a tourist on accommodation and entertainment in the area – a great deal depends on how and where that money is subsequently spent. If, say, 80 per cent of it is spent by the first recipient on *local* goods and services, and if the suppliers of those goods and services similarly spend 80 per cent of that 80 per cent locally, and so on, then eventually £5000 circulates locally, giving a 'multiplier' of five. But if just 20 per cent is spent locally at each round, then the total spend would be only £1250, a multiplier of 1.25. From that simple example, the consequences of the 'local multiplier' for local employment and for the general prosperity of the area become obvious as do the attractions of systematically trying to plug at least some of the 'leaks in the bucket'. As Douglas put it (1994: 11–12) 'the challenges of stemming or reducing the outflow of profits, incomes, sales and investment funds ... [and of] maximising the internal (re)circulation of dollars and import substitution are common objectives of community economic initiatives among Canadian communities'.

In that spirit, the New Economics Foundation (2001) is helping partnerships of local stakeholders in Britain to consider some or all of the following ways of 'plugging holes in the bucket', once they have researched financial flows into, within and out of their particular local economy:

- creating community-run businesses that deliberately buy *local* goods and services and reinvest any profits locally;
- attracting or developing businesses to fill identified 'holes in the bucket' – i.e. to supply some of the goods and services commonly bought from afar;
- persuading 'big spenders' in the local economy (e.g. local authorities) to source locally more of the goods and services they need – for example by favouring local contractors in awarding tenders;
- linked to that, providing occasions for small local businesses to 'meet the buyers' (such as those employed by the local authorities) and thereby to market more successfully their goods and services locally;
- setting up training programmes so that local people are better able to compete for jobs hitherto filled by in-commuters;
- promoting 'local food' initiatives – such as the farmers' markets referred to elsewhere in this chapter;
- promoting the local informal economy (e.g. via LETS schemes) so that local people needing particular goods or services are less likely to spend outside the area;
- reducing energy use, given that a very high proportion of gas, electricity and solid fuel normally comes from distant sources;
- providing services for in-commuters (e.g. childcare and catering) so that less of the salaries they receive leaks out to their place of residence; and
- encouraging the use of local shops and service providers.

The argument is that such import replacement and increased self-reliance can enable a town to develop sustainable commerce and to improve its competitive position vis-à-vis other areas – while simultaneously protecting or generating local jobs. Whether every locality can simultaneously and successfully do this is an intriguing question not wholly resolved in the relevant literature – but as a strategy for helping selected areas with particular problems to regenerate their local economies, 'plugging leaks' seems to offer real promise whilst respecting the precepts of sustainability detailed in this chapter.

The following case study develops many of these points further.

CASE STUDY 3: A MODEL SUSTAINABLE VILLAGE

In 1998, Dorset's Rural Community Council published an attractive eight-page pamphlet, written by Graham Duncan, portraying a hypothetical 'sustainable village' (Dorset Community Action, 1998). Its purpose is to encourage people living in villages and small towns to adopt a more sustainable lifestyle themselves and, by their example, to promote sustainable development at the local community level.

Two 'big ideas' of sustainability underlie the suggestions made: first, live off the earth's 'income' not its 'capital'; and, second, reduce your 'inputs' and re-use your 'outputs'. These messages are directed at both the individual and the local community together with a tacit third big idea as well, namely 'do all this locally'. All 16 of the subsequent ideas for sustainable development flowed from those principles and are listed below in four categories.

**Consume Local Produce**

- *Vegetable Box Schemes* offer a way of eating locally produced organic fruit and vegetables, create local employment and enable producers to avoid expensive distribution and marketing. Customers buy on a subscription basis and agree to take a box of vegetables each week and collect from a central pick-up point.
- *Food growing projects* are community owned and run and include community orchards, gardens and farms. Typically, the local community leases the land or orchard, often badly neglected, works it together and shares the harvest or sells the produce.
- *Local produce directories* are 'the green alternative to the local yellow pages', designed to help people obtain locally produced food and other produce. Criteria are needed to define what is 'local produce' and attention must be paid to updating the directory and to its marketing and distribution.
- *Local produce markets*, pioneered by the Women's Institute, are now multiplying with 'farmers markets', for example, proliferating since 1998. Often the traders buy a share in the venture and run it on a co-operative basis.

**Share resources and build communities**

- *Lift sharing* reduces the number of vehicles on the road, reduces pollution and energy use and can build friendships. Such schemes can be co-ordinated by employers or by local groups who act as 'brokers', matching drivers and potential paying passengers. The 1988 Road Traffic Act legalised this so long as no profit is made.

- *Community vehicles* are communally owned, with volunteer-operated community minibuses often running regular routes as well as being available for hire, and 'vehicle pools' involving sharing the purchase and running costs of a car, van or moped, perhaps among several people who book it for use when they need it.
- *Tool and machinery pools* involve a community group keeping a store of tools and machinery that is available for use by members of the pool on a subscription or fee basis. The tools can be bought by the group or lent on a long-term basis by members.
- *Local Exchange Trading Systems (LETS)* allow members of a community to both offer and purchase services and goods to/from one another using a local currency designed for the purpose. 'Granny sitting' and 'car repair', for example, are appropriately priced and the local currency obviates the need for direct barter.

**Reduce, re-use and recycle**

- *Recycling projects* involving the community can be of many types from simply collecting glass, paper, metals or textiles and selling them on, to running a recycling store or a community business which repairs and resells furniture, household goods, bicycles and other equipment.
- '*Green Lifestyles*' is a project to help people take 'do it yourself' environmental action; it encourages individual people to change, but with others in the same community working together for mutual support. There are 'action packs' to support such action in the home, the workplace or the school.
- *Composting schemes* generally involve the collection from people, or the delivery to a collection point, of household organic waste. In due course, the compost can be used by community farms or gardens or, if sold, can be the basis of a small community enterprise.
- *Ideas* can be recycled too, often with adaptation to local circumstances. In practice this means 'plugging in' to relevant networks. Increasingly, this can involve the Internet, but local clubs and societies can serve a similar function. Community facilitators can foster the revival or formation of such societies.

**Build the local economy**

- *Community enterprises* are trading ventures which are managed by local people for local people, with any benefits or profits retained by the organisation. They include co-operatives, development trusts and other community businesses. Many of the various activities listed in this case study can be pursued as 'community enterprises'.

- *Credit unions* are mutual non-profit organisations (one type of 'community enterprise') which enable communities to 'keep money local' and to save and borrow, at a reasonable rate of interest, through a 'bank' that they run themselves.
- '*Resource centres*' provide, in one building, a focus for training, information and access to IT. They generally provide computer and communication facilities, IT and other training, and a range of information such as on local job vacancies. They not only help individuals but they also provide a spur to local enterprise.
- *The promotion of local businesses* can take several forms – a widely available business directory, a regular 'flyer' in which businesses advertise their services and any current offers, or a campaign run by the local council or chamber of trade. The shops of a small town can be marketed as collectively rivalling an urban supermarket, with less travel involved.

All of these initiatives are small in themselves but, collectively, they comprise a manifesto for sustainable living at the local community level. They place equal emphasis on the environmental, economic and social benefits of an 'alternative local economy' and they seek to educate at the same time. But research is needed on:

- how far they and similar initiatives successfully reduce the flow of goods, services, money and people to and from the rest of the world; do they enhance in some small measure, the cause of local self-sufficiency?;
- how far they produce as a by-product a more vibrant and caring local community, and what the tangible benefits of that are;
- how far such one-off projects – 16 in the above listing – mesh together to provide synergy. (On the face of it, composting, community gardens, local produce directories and markets, and perhaps community enterprise and tool and equipment pools as well, collectively reinforce one another – but does this happen in practice?); and
- the best geographical scale for all this – how local or 'non-local' should the perspective be? Individual villages and small towns or the wider areas typically the domain of rural development programmes?

---

CASE STUDY 4: A DIRECTORY OF SUSTAINABLE RURAL INITIATIVES

---

This Directory, published in 1998 after two years' research, was jointly produced by Forum for the Future – a charity committed to building a sustainable way of life – and the Countryside and Community Research Unit of the University of Gloucestershire, with Phil Allies as the principal researcher (Cheltenham Observatory, 1998). The project's point of departure was an observation that throughout Europe millions of pounds are being invested in rural economies through initiatives designed to protect communities, jobs, landscapes, cultures and ecosystems. But how much of this money is actually advancing 'sustainable development' as a policy goal?

It therefore had two main objectives: the first was to develop and publicise a user-friendly *process of appraisal* which would lay bare the main negative and positive contributions to sustainability made by any actual or potential rural project or programme; the second was to publish neatly encapsulated *profiles of about 50 projects or programmes which were demonstrably sustainable* according to that appraisal process – not for blind replication elsewhere but as food for thought on what is possible.

The intended readership of the Directory was anyone involved in making decisions about rural development projects, programmes or policies such as local authorities, government agencies, local development partnerships, project promoters and community groups.

The research work involved four exercises:

1. developing the appraisal technique with reference to the theory and practice of sustainable development and careful consultation with potential users;
2. inviting a wide range of British and continental organisations and individuals involved in rural development to submit information on the projects or programmes in which they were involved – this exercise producing a 'long list' of over 500 initiatives for appraisal;
3. appraising those initiatives by using the appraisal technique – an exercise that deliberately involved those who were directly involved on the ground; and
4. selecting about 50 case studies for inclusion in the Directory and making it and a methodological note available on the Internet and, in summary form, as hard copy.

By way of an example, one of the 50 case studies of good practice included in the Directory was the SPARC local development partnership already reviewed as Case Study 2 in the present volume. It was appraised independently by both the researcher and the local SPARC manager using a series of indicators which were then conflated into nine sustainability criteria. These were essentially the 'goals for sustainable local development' already set out with minor modification in this chapter (see table 2.1). Thus SPARC scored particularly well on 'developing skills, awareness, knowledge, health and motivation of local people' and on 'maintaining and encouraging diverse cultures, traditions and activities', without generating a negative rating on any of the other seven criteria. The Directory also set out for each case study of good practice some brief descriptive information capable of capturing the interest of anyone dealing with similar issues. This information related to the project or programme's mission, focus, sector, beneficiaries, funding and plans, as well as to its difficulties and constraints and any lessons learned.

Thus the Cheltenham Observatory Directory of Sustainable Rural Initiatives proved useful in three respects: clarifying the key elements of sustainable development; translating them into an evaluative tool; and developing good practice in the presentation of good practice.

---

## NOTE

1. Mention should also be made of so-called 'quality of life capital' based on research conducted by the Countryside Agency, English Heritage, English Nature and the Environmental Agency (2001). This is not a 'fifth capital' – the four listed above remain totally comprehensive. Rather it is a *tool* to help planners and developers assess the 'quality of life implications' of possible developments, by, in effect, considering the four capitals from a different perspective. 'Consider a small mixed woodland on the edge of a town. "Quality of life capital" says it's not the hectares of woodland that matter in themselves, it's the capacity of the wood to provide tranquil recreation, a habitat for rare species, to stabilise the soil, retain water, mop up carbon dioxide and local air pollution – and perhaps also to support a livelihood in charcoal burning or coppice timber products' (p. 2.) The tool provides a way of assessing such benefits.

## SEE ALSO...

Without exception, all of the case studies in this book relate in some way to the local pursuit of sustainable development. Numbers 2, 9, 12, 20, 24 and 28 provide a good selection.

## SELECTED FURTHER READING

Bryden (1994), Selman (1996), Royal Society for the Protection of Birds *et al.* (1999). Also the UK government's 'quality of life capital' website *www.qualityoflifecapital.org.uk*. Specifically on economic matters see New Economics Foundation (2001) and the journal *Local Economy*. In addition *www.sustainableplace.co.uk* relates closely to Case Study 3.

# 3

# Innovation: Breaking the Mould

'Will you, won't you, will you, won't you, will you join the dance?'

Lewis Carroll, *Alice in Wonderland*

## DEFINITIONS AND IMPORTANCE

Taxibuses, dial-a-ride, telecottages, village design statements, ecomuseums, good neighbour schemes, green audits, village trusts, post offices in pubs, advice shops, train-and-build schemes, farmers' markets … all are examples of activities designed to help regenerate the local economy of particular rural areas and/or to improve the quality of life of the people living there; and all are typically introduced by local people and organisations as a response to some perceived problem unlikely to be solved by traditional ways of doing things. In short, all are 'innovations' at the moment of their introduction in some particular place, and this chapter is concerned with such innovations in a rural development context.

The topic is important because innovation is a key component of development, with doing 'more of the same' unlikely to drag a local area out of its social and economic malaise. If we take 'development' to mean a sustained and sustainable process of economic, social, cultural and environmental change designed to enhance the long-term well-being of the whole community – as suggested earlier, in Chapter 1 – and 'rurality' to denote areas of low population density containing scattered dwellings, hamlets, villages and small towns, and characterised by predominantly extensive forms of land use, then the adoption of new ways of doing things is central to the challenges implied. In short, rural development *requires* a mass of innovative decisions to be taken by individuals, households, firms and voluntary and public bodies in areas which are both denied the economies that accrue from large-scale operations or from geographical agglomeration, and are replete with problems relating to remoteness, rapid social change, service delivery, environmental conflict and a need to restructure rapidly the economic base.

So it is not surprising that many funding agencies now make their support of rural development programmes conditional upon an assurance that most of the projects to be promoted locally will, to some significant degree, be new or different from what has gone on before. For example, the European Union's LEADER I and LEADER II programmes were explicitly designed to 'support innovative, demonstrative and transferable measures that illustrate the new directions that rural development can take' (Commission of the European Communities, 1994a: 6). And the Commission insisted that 'the innovative nature [of each local programme] must not be confined to the method [of managing that programme] but must also be apparent in the technical content of the product, the production process, the market or some other aspect [of the projects supported]' (Commission of the European Communities, 1994b: 50).

That requirement was reiterated in the planning of the third LEADER programme running from 2000 to 2006 under the banner of LEADER Plus. As the EU's Agriculture and Rural Development Commissioner made clear, 'we need a kind of showcase for what we are trying to encourage on a larger scale in the mainstream rural development programmes; the emphasis of the new initiative, "LEADER Plus", should therefore be on supporting pilot rural activities...[it] must be a laboratory for rural development to encourage the emergence and testing of integrated and sustainable development approaches' (Fischler, 1998: 2).

The premium placed by development agencies on doing something different – on 'innovating' – is also apparent in the mass of 'Good Practice Guides' that now pour onto the desks of anyone concerned with rural development. These guides, in effect, proclaim 'Here are initiatives that have proved beneficial elsewhere, why don't you try them?' Most comprise attractively illustrated outlines of successful projects deemed worthy of wider application. They relate, for example, to village retailing, rural transport and childcare, to partnership working, to consensus building, to the marketing of quality tourism, to the execution of 'baseline studies' prior to planning a local area development programme and to participatory techniques. (These specific good practice guides are listed in the author's earlier review of innovation – Moseley, 2000a.)

But all of this official exhortation to adopt good practice tends to ignore a body of theory – 'innovation diffusion theory' – which could help considerably in making it more effective, and the rest of this chapter seeks to link that theory to the practice of achieving change on the ground. But, before that, we need some definitions.

Following Rogers' masterly review of the subject, we will define *innovation* as 'an idea, practice or object that is perceived as new by an individual or other unit of adoption. It matters little ... whether or not an idea is objectively new as measured by the lapse of time since its first use or discovery ... if the idea seems new to the individual, it is an innovation' Rogers (1995: 11). Thus innovation is not 'invention'; it is, as another commentator put it, in the context of rural development, 'doing something which did not exist before in a particular territory or technical area' (Vuarin and Rodriguez, 1994: 15).

As for *diffusion*, it is 'the process by which an innovation is communicated through certain channels over time among the members of a social system. It is a special type of communication in that the messages are concerned with new ideas' (Rogers, 1995: 5). And in this context a 'social system' is a set of individuals, informal groups or organisations sharing a common concern or set of goals and potentially capable of adopting the innovation in question.

Putting those two concepts together, *innovation diffusion theory* (strictly, the expression 'innovation diffusion *and adoption* theory' would be more accurate) is concerned to explain how it is that innovations, at first taken up by just one or a few 'adopters', come, over time, to be adopted (or indeed not adopted) by others in the relevant 'social system'.

## SOME KEY ISSUES

Before going on to consider how to promote successfully the adoption of innovations helpful to rural development, it is useful to address four key questions about the adoption of innovation generally – in whatever milieu:

First, what are the *typical characteristics of those innovations* that tend to be most quickly adopted? Based on his examination of over a thousand empirical studies of innovation, Rogers suggests that there are five.[1] An innovation is more likely to be adopted if it is

| | |
|---|---|
| observable | easily visible by would-be adopters |
| advantageous | perceived as distinctly more promising than the known alternatives |
| compatible | consistent with existing values, past experience and current needs |
| simple | easy to understand and to use |
| reversible | capable of introduction on a trial basis |

Second, what are the *typical characteristics of the 'early adopters'* – i.e. of those people, groups or institutions that innovate most readily? Paraphrasing Rogers and other commentators such as Ruttan (1996), there are essentially just three:

| | |
|---|---|
| Their business or professional circumstances | Early adopters tend to have a pressing problem, i.e. a clear gap between the actual and the desired. They also have access to the resources necessary to change their current practice. |
| Their personal attributes | Early adopters tend to have good formal education, relatively high socio-economic status, are socially mobile, open to new ideas and able to cope with risk and uncertainty. |

| | |
|---|---|
| Their connections | Early adopters tend to be exposed to a variety of media, to be well connected in interpersonal networks and to have good contacts with opinion leaders, change agents and other early adopters. |

Interestingly, the most recent research has tended to elevate the third of these to prime position – in short, to stress the overwhelming importance in innovation adoption of networks and networking. In other words, building and maintaining relationships with other actors is crucial. As Winter (1997) put it in relation to the variable readiness of British farmers to adopt innovation, the key need is to understand them as actors unevenly placed in the 'knowledge society'.

Third, what is *the typical sequence of events culminating in innovation adoption?* The research suggests that there are generally six stages, even if it is too simplistic to imagine a simple, linear progression in all cases:

1. the recognition of a problem or need, meaning some aspect of the status quo which needs to be changed;
2. the acquiring of knowledge of a relevant innovation and how it functions;

   (Interestingly, stages 1 and 2 are, in practice, frequently reversed – thus the recognition of 'need' can *follow* the learning of a way to assuage it, as any honest advertiser will confirm!)

3. the forming of a positive attitude towards the innovation;
4. the decision to adopt the innovation;
5. the implementation of the innovation; and
6. the consolidation or reinforcing of the innovation such that it is unlikely to be reversed or abandoned.

The fourth and often neglected question is *What are the typical consequences of innovation?* Here Rogers indicates a shortage of helpful research. In part this is because, to be useful, the research has obviously to be both long-term and capable of filtering out alternative causes of the 'consequence' in question. But also relevant is the frequent complacency of the 'change agencies' fostering the innovation, who tend to assume that the effects will inevitably be the positive ones they desire, and therefore see little need for subsequent monitoring and evaluation.

But many innovations have consequences that are unanticipated and/or undesirable. Indeed, some would be almost universally condemned as involving change for the worse – the spread of drug-taking among rural youth, for example, or the more extreme forms of factory farming. Other innovations, while apparently benign, often widen the gap that existed at the outset between the early and the late adopters – a case of 'first up, best dressed'. So we must beware of 'pro-innovation bias', of assuming that all innovations that benefit their immediate adopter bring net benefits to the wider society.

But whatever the precise causes and consequences, there is certainly a great deal of innovation occurring in rural Britain and further afield. The pace of local innovation will need to quicken, however, as the context within which rural societies and economies exist changes ever more rapidly. The continuing erosion of village and small-town services by the superstore and, increasingly, the Internet; the growth of second-car ownership; the crisis in agriculture and further reform of the Common Agricultural Policy; the growing reluctance of the state to pay for high-quality services to be delivered in small remote outlets; the absorption into popular culture of concern for the environment – these and other contextual changes will demand that new ways of doing things, in other words 'innovations', are developed and adopted quickly.

With that in mind, and given the mass of research that has been undertaken to date, it is unfortunate that we still know very little about how key rural, but non-agricultural, actors actually think and behave when it comes to countenancing possible change – shopkeepers, school governors, local councillors, village hall trustees and small-business people, for example. Much less do we have a clear view of the attitudes and behaviour of the 'stakeholders' that surround any potential project, such as housing associations and private developers, the health authorities, development agencies, local authorities, bus companies, breweries, banks, privatised utilities and the post office. Each of the latter crucially influences the ability of those 'key local actors' to innovate at the local level.

## THE TOOLKIT

Moving on to practicalities, how can the innovations necessary for rural development be best promoted on the ground? A starting point, surely, is provided by the three lists of characteristics set out earlier – of those innovations which are most readily adopted; of those people and institutions who are typically most ready to adopt; and of the sequence of steps that have normally to be negotiated if an innovation is to be adopted and subsequently consolidated.

That body of theory, coupled with the practical experience of a host of development agencies such as England's Countryside Agency and the former Development Board for Rural Wales, points to a range of possible measures which are briefly summarised below. This list may seem less abstract if it is considered with some real 'innovation diffusion' issues in mind – for example, the task of persuading small post office franchisees to embrace computerisation, or farmers to diversify into landscape management, or parish councils to produce a 'parish plan'. Relevant measures in this regard include:

- systematically gathering information on the personal and business circumstances and attitudes to change of the potential adopters. This may well suggest some priorities for action including an initial targeting of those who are most likely to adopt first and are well connected to other potential adopters;

- deploying 'field staff' (or 'facilitators' or 'business advisers' – the terms vary) to meet potential innovators, to discuss their problems and preferences and to offer information and advice;
- working with other stakeholders whose actions may affect decisions to innovate locally – for example, the banks which might put business through the post office network, or the breweries whose conservatism might discourage pub tenants from opening a post office inside the pub;
- encouraging 'early adopters' to converse with their peers, as it is often people who 'have been there' who are the best ambassadors of change. This may involve developing new machinery for the exchange of ideas, for example, countywide (and maybe 'virtual'?) 'village shopkeepers associations' at which new practices can be discussed informally and experience exchanged;
- identifying and supporting 'opinion leaders' – those people in a local area whose views are respected and whose example is often followed;
- providing opportunities for the sceptical to witness success on the ground. Here the example of the LEADER programme is relevant; funds were made available for the transnational exchange of experience, one example involving Greek farmers travelling to Scotland to see how bed and breakfast provision benefits their Scottish counterparts;
- funding pilot programmes, judiciously located and subsequently networked. An example is the funding of some young person 'foyers' where accommodation, training and personal support are all offered in the same complex of buildings;
- offering part-funding of the capital costs of adopting a new working practice. In the 1990s, for example, many Rural Development Programmes in England deployed 'business initiative funds' – no more than £1000 per successful applicant – to foster innovative practice. This was a tiny amount, but the condition that applicants had first to prepare a business plan and discuss it with a business adviser often led to an ongoing relationship between the would-be innovator and an appropriate 'facilitator', and to a much greater readiness to try something new;
- offering ongoing support or 'hand-holding' at critical phases before, during and after the adoption of the innovation. Many French training agencies offer what they call '*accompagnement*' – 'training' viewed not as 'sending people off on a course' but as working alongside groups of individuals as they grapple with introducing something new in their workplace (see Case Study 26); and
- finally, and recalling the point that innovation often occurs first where an innovative culture has already developed, geographically or sectorally targeted training/awareness/education programmes may be needed to get things moving in 'backwater' areas.

It is not difficult to see how many or all of these suggested measures could be incorporated into a 'strategy for rural innovation' tailor-made for a particular area or sector of a local economy or society. Several are demonstrated in our consideration, below, of two innovations relating to the process of rural development in Britain. The first is the parish appraisal, a community development innovation that has now diffused widely across rural Britain. The second is an innovation 'model' with great scope for local adoption and adaptation, namely the joint provision of services in a single building or vehicle.

Stemming from the earlier theoretical discussion, four key questions will underpin this consideration of case studies:

- first, how might we recognise and choose innovations that are most appropriate in the promotion of rural development?;
- second, what factors aid the swift diffusion and early adoption of innovations for rural development?;
- third, what are the typical consequences of innovation for rural development?; and
- fourth, how can the subsequent spread of innovation be best promoted?

## CASE STUDY 5: THE PARISH APPRAISAL AS AN INNOVATION

As Chapter 10 and Case Study 19 will emphasise, parish or village appraisals are questionnaire surveys *of*, *by* and *for* the local community, undertaken to identify local characteristics, problems, needs, threats, strengths and opportunities. They are a means of taking stock of the community and of creating a sound foundation of awareness and understanding on which to base future action. The three key words in the definition are 'of, by and for' and the most important of these is 'by'. In other words, parish or village appraisals are 'do it yourself' exercises in community development, and since in the vast majority of cases they have been carried out in a particular local community for the very first time, they have been 'innovations' several hundred times over.

Indeed, it is for that very reason, the fact that appraisals have been around for a long time, that they are highly appropriate to this consideration of how innovation spreads. Ironically, focusing on something much more modern, such as the 'parish plans' being introduced by the Countryside Agency from 2001, would throw much less light on the process of 'innovation'.

But just how many British rural communities have undertaken 'parish appraisals' is impossible to say. The first was probably Stocksfield in Northumberland in 1971–2, the Town and Country Planning Act 1968 having introduced the then novel requirement that local opinion be canvassed as part of the plan-making process and the people of Stocksfield being keen to collate and express their opinions (Lumb, 1990). Thereafter, the innovation spread with gathering momentum; by the late 1980s around 400 appraisals had been undertaken in rural England and a decade later the figure had risen to over 2000, taking England and Wales together (Moseley, 1997). By today we can probably say that appraisals have been undertaken of, by and for at least a quarter of our local rural communities.

Typically they have been managed, over a period of about two years, from start to finish, by *ad hoc* 'parish appraisal committees' comprising a dozen or so people, most of them already active in such bodies as the parish council, the village hall committee and the Women's Institute, and/or in local organisations such as playgroups, youth clubs or village societies where these already exist.

To explore in more detail the spread and the impact of parish appraisals, the present author examined the experience of 44 parishes in Gloucestershire and Oxfordshire (the 'South Midlands') where appraisals had been carried out in the 1990–94 period (Moseley *et al.*, 1996). Here we focus explicitly on those 44 appraisal exercises construed as 'innovations'.

The 44 were a sample of the 90 or so appraisals undertaken in the two counties in that five-year period. The interesting question is this: 'why *those* 90? and not another set of 90, or for that matter 190, of the 574 parishes which make up the more rural parts of those two counties?' It is that question that intrigues the student of innovation diffusion. Was it something to do with the social structure of the parishes in question or to do with their particular circumstances at the time? Or was it to do with the behaviour of external agencies such as the Gloucestershire and Oxfordshire Rural Community Councils, or to some other factor?

From the evidence of the South Midlands study, it seems that four elements needed to be in place for the 'parish appraisal' innovation to be adopted:

- a concern felt by at least one potentially influential resident that 'all is not well in this parish' – very few appraisals were carried out from sheer disinterested curiosity;
- a 'trigger event', which could have been an opportunity or, more commonly, a perceived threat. The most common 'triggers' in the South Midlands were announcements that a Local Plan was to be prepared by the local authority. Others included the worrying prospect of major local developments such as a bypass, holiday village, gravel pit or estate of speculative housing, and a particularly disappointing turnout at a parish council election;
- an offer of moral or technical support, most notably by an officer of the Rural Community Council who could point to successful appraisals in the same county and be reassuring about their execution; and

**TABLE 3.1** ***The implementation of parish appraisal recommendations and action points; evidence from the South Midlands (after Moseley et al., 1996)***

| Key factors influencing the implementation of the parish appraisal's recommendations and action points | per cent of parishes where this was clearly the case |
|---|---|
| One or more dynamic and motivated individuals in the parish determined to achieve success | 91 |
| Explicit attention in the appraisal report, or soon afterwards, to the process of follow-up | 68 |
| Enthusiastic endorsement of the recommendations by the parish council | 50 |
| Clear 'legitimation' of the report and its recommendations by local consultation/consensus | 43 |
| Realism – i.e. recommendations broadly consistent with political realities and resource availability | 41 |

- the subsequent availability in the parish of at least half a dozen energetic people ready and willing to come together and take the idea forward.

It cannot be said that all four elements *had* to be present for the appraisal to 'happen', though normally they were. Nor, of course, can it be said that they were never present in the non-adopting villages.

With our focus still very much on the adoption of innovation by rural communities, there is one other clear way in which parish appraisals are interesting. Did they lead to *further* innovation? If so, how and why? Again, the South Midlands study gives some clues. The 44 appraisal exercises under scrutiny had collectively generated some 420 'recommendations and action points' based on the concerns expressed by local residents in the surveys. These ranged from 'We must do something to curb traffic speeding through the village' through 'Something must be done to retain or improve the village school' to 'Let's plant some trees in such and such an area of the village'. (see Case Study 19 for more detail on this).

Many of the projects that subsequently resulted from these appraisals cannot reasonably be construed as 'innovations' since they resembled or built upon earlier initiatives in the parish in question. But some certainly were. Examples include the launch of recycling schemes, good neighbour schemes and community transport initiatives of one sort or another, and it is interesting that the most successful local innovations generally had one particular characteristic in common – they had been capable of being launched by the local people themselves with relatively little help from outside agencies. That is not to say that the construction of some affordable housing or the installation of traffic-calming schemes on busy through roads – each very largely the province of some 'superior' agency – never resulted, but, if they did, they were the product of years of persistent lobbying.

Finally, it is interesting to note (Table 3.1) that a high proportion of a parish's 'recommendations and action points' tended to be turned into successful action if some or all of five key factors were in place relating to the innovation, the potential innovators and the socio-political context surrounding them.

## CASE STUDY 6: THE JOINT PROVISION OF DISPARATE RURAL SERVICES

One particular response to the increasing difficulty of providing services for rural communities has been to locate two or more distinct services within a single building or vehicle. This both reduces the costs which fall upon one or both providers and increases the scope for 'multi-purpose trips' by the consumer.

This is not new of course; in their different ways both the village shop-cum-post office and the multi-purpose village hall provide very British, and by now long established, examples of this principle. But surveys in 1996–7 and 2000 indicated the wide range of services now delivered in this way and generated advice on how good practice might best be spread (Moseley and Parker, 1998; Moseley *et al.*, 2000). In the language of this chapter, the task in that research was to appraise a set of innovations, assess their longevity (hence the follow-up survey) and advise on the facilitation of their adoption.

The term 'tandem operation' was coined for the phenomenon in question; this was defined as 'the provision of separate and distinct services by two or more agencies, in premises or a vehicle provided by one of them or a third party'. As there was and is no national or even county-level database of such 'tandem operations' the research involved visiting three rather different rural counties – Norfolk, Nottinghamshire and Cumbria – and seeking out examples of them with the assistance of each county's Rural Community Council.

To illustrate the variety of practice discovered, Table 3.2 gives some basic information on just two 'tandem operations' in each of the three counties. In each case, one relates to a building and one to a vehicle. All of the 'tandem operations' examined were clearly innovative, i.e. there was nothing similar to be found in the vicinity. And while each demonstrated the same disadvantage, namely some compromise on service quality when compared with conventional 'single provision outlets', they also shared the same advantage – cost savings arising from sharing the building, vehicle and sometimes personnel of another service provider.

**TABLE 3.2** ***Some rural 'tandem operations' in three counties (from Moseley and Parker, 1998)***

| An innovative local 'outlet' | The services delivered therein | The 'operational partners' |
|---|---|---|
| **Norfolk** | | |
| Stoke Ferry School | primary school, church, community centre/ theatre, GP surgery | Stoke Ferry community association, Norfolk County Council, head teacher, Church of England, local GP |
| The Broadland Information Bus | information within the remit of the various partners, plus payment of council rent and tax | Broadland District Council, local citizen's advice bureau, Age Concern, Inland Revenue |
| **Nottinghamshire** | | |
| Collingham medical practice and library | GP service with a volunteer car scheme, dentistry, chiropody, dispensary and other health services, public library | GP fund-holding practice, other health professionals, Nottinghamshire County Council (library service), Volunteer car service organiser |
| A milk delivery vehicle-cum-informal 'cash point' | delivery of milk, cheque-cashing service | milk delivery franchisee, dairy company (which turned a 'blind eye') |
| **Cumbria** | | |
| Private residence, post office and part-time surgery in Roadhead | post office, GP surgery | local GP, Post Office Counters, resident acting as postmistress, two health authorities |
| Coniston Institute | village hall, day centre for elderly, learning disability day centre, plus (now discontinued) citizen's advice bureau | Coniston Institute committee, local trust, citizen's advice bureau, Age Concern, Cumbria County Council, community health trust, Coniston parish council |

But the research revealed three further benefits that frequently accrued First, an increase in the number of potential clients available to one service provider on account of their arriving to partake of another – for example, customers coming into a combined pub/post office to buy some stamps and staying to have a pint. Second, synergy, with one service provider benefiting from close proximity to another – for example, health and social work professionals learning from each other inside a multi-purpose village centre. Third, increased social interaction as some of the tandem outlets became, in effect, 'community fora' or village meeting places, village halls accommodating post offices exemplifying this quite well.

To summarise, considerable advantages were apparent in this emerging pattern of rural service delivery, especially in areas of low population density where, increasingly, it was a case of 'either we hang together or else we hang separately'. Continental experience served to reinforce this view – for example, the French government's 'One Thousand Villages' programme designed to promote *multiples rurals* in communes where the village's last outlet, be it a general store, boulangerie, café, or whatever, was facing extinction. The government's main conditions for putting money into such schemes was that the entrepreneur in question had to produce a business plan justifying the services to be provided, with reference to some market research of the needs of the local community, and that the commune council had also to pledge its support and commitment.

Looking at all of these 'tandem operations' in the light of 'innovation diffusion theory' it is clear that a common element is generally the existence of an enterprising individual at the local level, often faced with a problem that precludes the 'do nothing' option and able to cope with risk and uncertainty. But, in addition, he or she needs to exist within a context of at least benign, if not positively supportive, 'higher level' agencies ready at least to countenance the proposed innovation. If those agencies are genuinely supportive and provide technical, financial or other assistance, then the innovation clearly has a much better chance of taking root and prospering.

Interestingly the 2000 follow-up survey of the 24 'tandem outlets' studied in 1997, discovered a picture of continuity and incremental change over the intervening three years with the sole exception of the closure of three post offices in the various outlets. Key factors in the change and continuity, affecting the 'bedding down' or else disruption of innovation, were the personal qualities of the outlet manager or proprietor and the policies and attitudes of the 'non-local stakeholders', such as the breweries, Post Office Counters and the health authorities.

However, in neither 1997 nor 2000 did we study the *absence* of 'tandem outlets' and perhaps the interesting question is not 'Why are there now a couple of hundred or so post offices located in pubs?' but 'Why are there not a lot more?' To answer that question we would need to delve more deeply into the various constraints impeding innovation – indeed into the attitudes and behaviour of innovators, non-innovators and the decision-makers surrounding them who may have a vested interest in maintaining the status quo or else may be ignorant of new possibilities.

Looking at both innovations, parish appraisals and 'tandem service outlets', some important conclusions emerge. Each attacks a constraint that is holding back positive change – respectively the lack of local machinery that would encourage consensus rather than division, and a lack of economies of scale in service provision. Each also 'builds local capacity' – be it in individuals who gain confidence and skills and a readiness to innovate again, or in groups and organisations. And each is realistic in the sense of being capable of implementation by local actors with only modest recourse to outside support. Finally, as with all the best innovations, each demonstrates subsequent adoption or adaptation by neighbours with similar concerns.

And what factors aided the swift diffusion and early adoption of these innovations? In each example the early innovators tended to be communities or businesses with dynamic leadership, well connected to sources of ideas, inspiration and advice and operating in a supportive institutional environment. But those conditions, while generally necessary, were rarely sufficient. Often there had to be a 'trigger event' coming on top of an earlier realisation that 'something is wrong'. That trigger event could be a perceived threat or danger, such as the closure of a post office or an announcement that a Local Plan was to be prepared, or it could be a new opportunity, such as the prospect of some LEADER funding. Finally, swift diffusion was fostered by a climate of external encouragement; commercial and community entrepreneurs may well be the motors of rural development but they flourish best if their efforts attract the support of outside agencies.

### NOTE

1. The reader may care to check out the applicability of this list, and of those that follow, in relation to some recent innovation in his/her own life – buying a DVD player, perhaps, or joining a health and fitness club to fight off the onset of middle age!

### SEE ALSO...

All of the other case studies in the book relate in some way to 'breaking the mould', to the introduction or consolidation of innovation. Numbers 1, 7, 9 and 28 provide a range of perspectives. Case Study 16 reports a particular service delivery innovation and number 19 says more about parish appraisals.

### SELECTED FURTHER READING

Rogers (1995) remains the 'bible' on innovation in social systems. Moseley (2000a) covers much of the ground of this chapter but in more detail. LEADER Observatory (1994a) and LEADER Observatory (1997b) consider innovation as a feature of the LEADER programme.

# 4

# Adding Value: Building on What's There

> Lord, thou deliveredst unto me five talents: behold, I have gained beside them five talents more.
> His lord said unto him, Well done, thou good and faithful servant ...
>
> Matthew 25. 20–1

## DEFINITIONS AND IMPORTANCE

If a farmer decides no longer to sell his crop of cherries directly for final consumption or to a distant food processing company, but instead to invest, perhaps with others in the vicinity, in the plant necessary to transform the cherries into cherry brandy and then to sell the brandy rather than the unprocessed fruit (as in Spain's Jerte Valley – see Case Study 7), he is 'adding value' to a local resource and thereby generating wealth and employment on or close to his farm rather than in a region far away. If he opens a shop on his farm or in the nearby town to sell the brandy to tourists at retail rather than wholesale prices, perhaps offering conducted tours of his orchards and the distillery for a small additional fee, he is adding still more value to his resource. This concept – what the French call 'valorisation' – is at the heart of 'endogenous development' – development based predominantly on local resources rather than on those imported from outside.

In fact the term 'valorisation' has a slightly broader meaning than the narrow 'adding of value' to a product which, duly enhanced, is sold at a higher price. It also tends to mean enhancing the value of a local resource even if it is not then marketed in the usual way. Thus a ruined castle can be 'valorised' by restoration, redundant unskilled labour by training or an underachieving institution, such as the typical English parish council, by a programme of assistance, empowerment and encouragement. In each case a local resource which originally had little or no value in economic terms becomes a real player in the process of sustainable development – 'sustainable' in that its exploitation occurs in a way that does not destroy future options for its deployment.

So our definition of *adding value* will embrace this rather larger conception: 'adding value is the sustainable exploitation of a hitherto underused local resource such that it generates wealth and employment in that area'. As such the notion is very close to that of 'enhancing an area's territorial competitiveness' as outlined in a range of recent documentation distilling the lessons of the LEADER experience across Europe (LEADER Observatory, 1999a, 2000a, 2000b, 2000c, 2000d), a key one being that those responsible for the fortunes of local areas must draw up their own 'territorial project', based upon undervalued local resources, and seek to market their products to a largely non-local clientele.

And it should be noted that the local pay-off from adding value to local resources is not just the creation of jobs there rather than elsewhere and the local economic multiplier that comes from the enlarged workforce spending at least some of its money in the vicinity (a concept already discussed in Chapter 2). Experience shows that additionally, and in various curious ways, the reappraisal and re-valuing of a local resource, hitherto idle or neglected, tends to have a wider social and cultural effect – increasing or reasserting a sense of local identity and pride of place that itself becomes a latent resource to be exploited in a continuing process of development.

## SOME KEY ISSUES

An important aspect of the adding of value to local resources concerns the *championing of local distinctiveness*. This line of thinking owes much to the growing concern in rural development circles about the effects of globalisation, as explained in Chapter 1. Increasingly, capital can be assembled wherever in the world production costs are lowest, and increasingly there is global penetration of what were once fairly secure local markets. In short, local economies are now effectively open to the cold blast of competition from producers thousands of miles away. In our cherry brandy example the real spur to change had been the flooding of established western European markets for fresh cherries by cheap imports from eastern Europe arising from a new trade agreement with some former communist states.

The key lesson from all of this is that, increasingly, local producers have to produce and market something a little different – something 'differentiated' from the competition – and this requires ingenuity both in reappraising the local resource base with a view to exploiting any distinctiveness and in adding value to those resources in a way that will please an increasingly discriminating clientele. The development strategy of 'adding value to local resources' requires a positive attitude both to the potential of local resources *and* to the implications of globalisation. In the latter case it means seeing the opening up of world markets as an opportunity as well as a threat, and seeking not a rejection of globalisation but a judicious positioning within it. In short, local development is increasingly driven by those who see the bottle half full rather than half empty, and the challenge for the champion of local development is to help bring about such attitudinal change and to back it in a number of positive and practical ways.

Fortunately, there is a host of factors that serve to support the strategy of championing local distinctiveness. To begin with, Europe's rural areas are extremely diverse in terms of their history, culture, landscape, environment and human and physical resources so there is certainly the basis there for product differentiation. Second, there is increasing consumer demand for local or regional quality products and experiences, including rural, cultural and ecotourism which, conveniently, *has* to be consumed on site. Third, the Internet and other ICT developments have transformed the ability of small producers to market their products to a global clientele. Fourth, the gradual liberalisation of Europe's labour markets and the end of 'job for life' expectations, have provided both an incentive for local economic innovation and a means of achieving it, with, for example, an increasing number of people combining a part-time job with self-employment. Fifth, and linked to that, is the growing importance of small and medium enterprises in innovation and job creation, partly again as a result of their ability to be flexible in a climate of rapid change. And finally, there is now a wealth of technical and other support available to the would-be rural entrepreneur – from European, national, regional and local sources. In short, the time is ripe for innovation and enterprise designed to add value to local resources.

In this regard, *the resource audit* is of major importance. Several of the examples of 'adding value' listed in Case Study 8 are clearly the product not of formal resource appraisal but, rather, of hunch or circumstance. But others did grow out of a more systematic study of the area, a conscious search for resources that were unused, underused, neglected, abandoned, forgotten or just poorly exploited. A taxonomy of such potential resources may be useful, and the reader is also directed to Chapters 2 and 11 for other perspectives on this.

- *Location*. Every place has a unique location with its own set of spatial relationships with other places, and these relationships develop over time as the provision of transport and communications itself changes. A roadside location suggests scope for farm shops selling produce direct to the passing consumer. More generally, proximity to large towns suggests other ways of interesting an urban market, whether as day-trippers and weekend tourists, second-home owners or 'pick your own' gatherers of fresh fruit and vegetables. Several of the case study examples demonstrate this. Conversely, remoteness – especially when it is coupled as it is today with easy access via the Internet – is its own resource, especially for those wishing peace, tranquillity and a wilderness experience.

- *Physical environment.* Clearly landscape, terrain, climate, water, energy sources, fauna and flora and interesting habitats all offer economic potential in different ways. In Britain, for example, it is clear that an attractive environment is a key factor in attracting not just tourists but also entrepreneurs and highly qualified personnel who first seek somewhere congenial for themselves and their families to live, and only then a source of employment or premises for the commercialisation of a business idea. Where new firms are born often says as much about residential preferences as about business considerations, an observation that has enormous implications for rural development.
- *Cultural heritage.* A purist would argue that 'landscape' should appear here rather than as part of the 'physical environment' as throughout most of Europe it is as much the product of men and women as of the forces of nature. Other elements of an area's 'cultural heritage' include buildings, factories and mines, redundant railways, local history and traditions, dialect, music, art, cuisine, festivals and folklore, traditional know-how, as in a distinctive handicraft industry, and the image and reputation of the area, as with the cachet of Cambridge, which has brought hi-tech firms not just to that city but also to surrounding villages. All of this provides something on which to build a development programme and it does so in two ways: first, cultural heritage can be marketed directly to the tourist, visitor and would-be resident or entrepreneur; second, it can be exploited as a way of stimulating local 'pride of place' and a sense of shared destiny, both of them frequent precursors of further development. (On this point see Ray on the rural 'culture economy' – the 'exploitation of a territorial identity through local cultural resources' (Ray, 1997: 1).)
- *Human capital.* Here we are concerned with the numbers, skills, education, health, attitudes, vitality, and capacity for risk-taking and for leadership of the population of an area. Positive attitudes to change and specific skills, perhaps made available by the closure of a large employer, or the product of an influx of migrants, may provide the basis for a new economic activity. Conversely, negative attitudes and the absence of a pool of marketable skills, coupled perhaps with a 'dependency culture' or expectation that the state will provide, produces a different challenge – that of 'adding value to the human resource' before it can add value to other resources.
- *Local institutions and social capital.* 'Social capital' determines the ability of local people and institutions to take joint, effective action at the local level. Existing clubs, co-operatives, extended families, trade unions, business associations, chambers of commerce and political parties, as well as the formal institutions of government, including the various local authorities, are also a resource for development especially as the lesson of so much of rural development is that it is a *collective* process of innovation and change. Most important perhaps are existing firms: do they have ideas for

new products or practices; are they open to new ideas, and do they have the culture and machinery to exploit them?
- *External relations*. The people and institutions of an area will already have contacts, relationships and perhaps markets and sources of capital outside the area. For them to have those contacts and relations is itself a local resource. Can they provide a basis for new economic and commercial initiatives?

If the above is a listing of *what* the local resource might comprise, the next questions are *who* might add value to them and *where*? As for 'who', the lesson seems to be that some alliance of individual firms and entrepreneurs, who must eventually shoulder the risk of innovation, and one or more public agencies offering technical support is usually most effective. Certainly that is the message of most of the LEADER examples considered briefly below. As for 'where?', the theory of local development requires only that the adding of value occurs within the area in question; whether it occurs on the particular site (often the farm) of the entrepreneur, or in a neighbouring village or small town where the necessary infrastructure is more likely to be available, is less important.

## THE TOOLKIT

There are two 'how to do it?' questions to be considered: how to identify the business opportunity from the long list of local underused resources; and how actually to add the all-important value locally.

As for *how to identify the business opportunity*, this comes close to matters of entrepreneurship and business strategy which are the subject of the next chapter. But two processes may be considered here, each requiring a leap of imagination or a capacity to see familiar things from a new perspective.

The first involves some sort of SWOT analysis (strengths, weaknesses, opportunities and threats – see Chapter 11) which may best be undertaken by a wide cross-section of the local community. A nice example of this is the brainstorming undertaken by children in ten primary schools in the county of Tipperary in Ireland, when the local LEADER group resolved to boost the culture of entrepreneurship in the area by backing small school-based businesses (see Case Study 10). More conventionally, a development agency might conduct an area-wide resource audit and SWOT analysis by involving a broadly representative range of informed local people, and then publicise the findings together with a call for project proposals from anyone with a related business proposition.

A second approach to identifying business opportunities based on underused local resources involves looking at existing value-added chains, establishing just where value is added to a local product as it moves from production to final consumption and exploring why it is that most value addition occurs outside the local area. Is it due to some overriding factor of economic geography or to a lack of the local collaboration between small-scale producers which would achieve economies of scale, a lack of venture capital, a lack of confidence in the ability to secure subsequent markets, restrictive land-use planning regulations, or what? Thus enlightened, consideration of the practicability of adding more value locally can proceed.

The second 'how?' question is *how to add that value locally*? What action needs to be taken and by whom? Again this is a big question and at least four other chapters of this book have something to say about it – those concerned with the introduction and fostering of innovation (Chapter 3), entrepreneurship (Chapter 5), the role of local partnerships (Chapter 9) and capacity building and community involvement (Chapter 10).

Clearly, the role of the development agency is crucial – successful innovation in a rural development context is rarely an individual exercise. Particular tasks for the agency (or agencies) may include:

- building up the area's identity and distinctiveness and helping the local project promoter to extract value from that identity/distinctiveness, for example by introducing a 'quality label' and making it available for use only by local firms passing a quality test;
- creating new collective structures aimed at making risk-taking more acceptable and more likely to succeed, for example associations of small and medium enterprises in the same business;
- researching the external context – the market for the proposed products, the likely competition, the sources of external assistance – and better positioning the local area in that context; and, going on from this, setting up marketing consortia and generally promoting the area, its businesses and its products;
- providing, or part-funding, the provision of appropriate training for management and for the workforce;
- making available some initial capital, whether as loans or as grants, for example for feasibility studies or to meet certain start-up costs; and
- more generally emphasising and supporting strategies for quality assurance: all the recent literature on rural development strategies built around adding new value to traditional resources underscores the need to produce goods and services of quality and distinctiveness. In contrast, mass production for a mass market is more properly the domain of large units, probably in the larger centres of population.

We move on now to consider a number of specific and recent cases of business development based on adding value to a hitherto unused or under used local resource. These case studies confirm a number of key conclusions to the preceding text, particularly the importance of partnerships involving local entrepreneurs and support agencies coming together in a common cause.

CASE STUDY 7: FROM FRUIT AND BERRIES TO WINE AND BRANDY

From Finland (LEADER Observatory, 2000e) and Spain (LEADER Observatory, 1995a) come two examples of developing a new use and a new market for fruit which hitherto had a limited or threatened market.

**Fruit and berry wine: the Savo region of Finland**

In 1993 the Muurevesi College of Agriculture and Horticulture in central Finland launched a training programme for small-scale fruit and berry wine producers. The course was designed to facilitate the emergence of a totally new industry in Finland, the organisers successfully anticipating a change in legislation which would allow the commercial production of wine on farms. The training provided an overall understanding of on-farm fruit and berry wine production and of how to develop this activity into a thriving business. The target group comprised farmers already involved in fruit and berry production and others planning to start rural enterprises. The project arose from a number of factors:

- Finland's latitude and its long, light summer days are very conducive to the growth of high-quality fruits and berries;
- in the early 1990s many regional and local development projects were launched in Finland to exploit this abundant raw material base by setting up new enterprises;
- Muurevesi College had for some time been arranging courses in non-commercial home-made wine production, and with impending legislation to legalise production commercially an opportunity was identified;
- the principal of the college had familiarised himself with wine production on various trips to western Europe and considered wine production on Finnish farms to be feasible; and
- the college had some relevant in-house expertise and the necessary international contacts to recruit further experts as needed.

The basic training lasted one year and combined theoretical and practical work at the college, projects undertaken at the participants' own enterprises, periods of practical work experience abroad and participation in trade fairs in Europe. It was organised so as to avoid the summer growing period and to allow the trainees to continue in full-time employment. Most funding came from the Kuopio County Government and the Ministry of Education.

By early 1998, when the author visited the Savo region, 33 Finnish fruit and berry wine entrepreneurs had been given licences to produce and sell their wines and a total of 38 persons engaged in these companies had received their training at Muurevesi College.

At the college itself, seven people were employed in 1998 as part-time teachers or in other activities supporting the new fruit and berry wine industry, and other developments have occurred as a direct result of this initiative. These include: a vocational degree course in wine production launched in 1997; special courses, and a consultancy service, provided for existing wine berry entrepreneurs and other professionals; and links established with three vocational colleges in northern Finland to deliver special training courses in berry wine production there. The Finnish Research and Information Centre for Fruit Wines was established at the college in 1995 and by 1998 employed three people in a purpose-built laboratory. Its recent work had included research into particular berries such as flavonoids, and a development project for distillates and other low alcohol products. In addition, wine production in the Savo area was beginning to stimulate rural tourism there.

Thus this has been a successful example of training playing a key role in the emergence of a completely new business based on local resources. It is also clear that other projects and income-generating activities have spun off from the initiative. Of central importance has been the good level of collaboration between the college, the local authorities, the local farmers and other business people, and an early recognition of the need to bring in specialist know-how from abroad.

**Cherry brandy: the Jerte Valley in the Estremadura region of Spain**

For many years the Jerte Valley, its flanks covered by thousands of cherry trees, enjoyed a near monopoly of cherry production in Spain. Its annual output of some 16,000 tonnes represented, and still represents, the livelihood of 3500 family farms, and each village has its small co-operative. About 80 per cent of the people of the Jerte Valley LEADER area depend on the cherry crop, but the early 1990s saw growing competition from other Spanish regions and from abroad. This competition forced the producers to be more professional in their production and marketing, to diversify into other cherry-based products and to commercialise other local products hitherto neglected, such as pears, figs, blackberries, raspberries and olives.

---

As the head of the local LEADER action group said in 1992, 'we must start from what we have – fresh fruit, the expertise of our farming population and our already highly efficient marketing network. But we cannot content ourselves with being simple suppliers of the raw material; we need to process our products on site, diversify them, offer a range' (LEADER Observatory, 1995a: 51–2). But though based firmly on local know-how, new technologies and new management and marketing systems have had to be developed.

It was in those ways that the local LEADER programme offered support – helping to develop new European markets for the new products, offering training for the fruit producers in the rationalisation of production, part-funding the development of new fruit preservation techniques and participating in the creation of a nursery laboratory for the development of new forms of cherry which would help producers to stagger production over a longer period and offer more choice to the consumer.

Also in 1992, the Agrupacion de Cooperaivas (a federation of local co-operatives) launched into fruit processing and LEADER part-financed some sophisticated quality control equipment. By 1995, 2000 tonnes of cherries that previously went unused were distilled to produce excellent kirsch – 200,000 bottles a year. Today a variety of cherry, plum, raspberry and blackberry brandies are marketed under the label Valle de Jerte.

Moreover, opening a marketing branch of the co-operative in Madrid helped the producers of other farm products in the valley, such as ham produced from mountain-raised, acorn-fed pigs. Such producers, however, had to respect a quality standard developed in the framework of the LEADER programme. Poor quality in relation to one product could unfavourably affect other products with the same local label.

The whole venture points up a number of lessons regarding decisions to add value locally to a traditional crop by developing new products based upon it. Most important is the multi-faceted nature of the whole exercise and the need for continuing support in resolving many of the issues that arise – new production technology, training in new production and marketing processes, developing marketing networks, help with quality assurance and the launching and managing of a 'label' for local producers of quality. But after several difficult years the net result has been that the economy of the whole valley now has a promising future rather than one curtailed by the ravages of globalisation.

## CASE STUDY 8: TEN FURTHER EXAMPLES OF ADDING VALUE LOCALLY

The following brief listing of ten further examples of the adding of value to hitherto neglected local resources is designed simply to show the breadth of what is possible. Each reported initiative figured in one or other of the local development programmes championed by LEADER I and/or LEADER II between 1991 and 2000. They come from different corners of Europe and are all a testament to rural diversity and to local inventiveness. Most of the evidence is drawn from LEADER Observatory publications (notably 1999g; 2000 a, b, c) plus, in the case of numbers 9 and 10, RSPB *et al.* (1999) and UK LEADER II Network (2000) respectively.

1. **Traditional know-how in Portugal**. In the Sousa Valley, in north-east Portugal, most women in 1990 had no source of income other than

poorly paid handicraft activities such as embroidery. The LEADER group made reviving the embroidery sector a central plank of its regeneration strategy, eventually handing over this task to an association, Caso do Risco (the 'fabric design centre'). This work has involved reintroducing traditional techniques, modernising patterns and computerising parts of the process. It has also involved training activities and promotional campaigns in high-quality markets. Between 1996 and 1999 the average pay of the women involved increased by 33 per cent and they now feel motivated and valued thanks to the technical support and professional recognition.

2. **Language and culture in west Wales**. The south Gwynedd area in west Wales is characterised by a strong sense of belonging to the Celtic culture and a shared desire to be different from neighbouring England. The LEADER group accordingly promoted several initiatives linked to promoting the Welsh language and culture and asserting the value of bilingualism. This approach sparked the local community's interest and commitment and also opened up several new development possibilities based on tourism linked to the literature, music, cultural heritage, architecture and historic sites of the area.
3. **Stone working in Tuscany**. A LEADER group serving the district of Firenzuoa, near Florence, has implemented an initiative to exploit the potential of a particular Miocene sandstone. The collaboration of the district council, associations of local businesses and individual firms made it possible to create new products using the sandstone. People now understand better the potential of this raw material, and this has resulted in product diversification, a significant increase in jobs and the widespread promotion of the area. Eleven small and medium-sized firms were created or expanded, employing 174 people, with a similar number of outsourced businesses and jobs. By 1999 there were two associations serving the industry; one supplies explosives and equipment and provides services to the businesses involved, including negotiating with institutions and other businesses. The second specialises in the restoration of historic centres using the stone.
4. **Second-home owners in Abruzzes, Italy**. The LEADER group in this case was concerned with local people's lack of self-esteem and their cultural dependence on the Rome metropolitan area. The aim was to devise initiatives that would enhance the value of the local cultural heritage and also sustain various social and health services which were at risk. One project sought to 'add value' to the influx of mainly weekend tourists from Rome and Naples, many of whom had second homes in the area but who contributed little to the local economy by not paying council taxes or in other ways. A programme was launched to encourage second-home owners to increase by 50 per cent the days spent each year in the area and thereby to spend more money there.

5. **Arts and crafts in the former East Germany**. In an area of Thuringia, badly hit by unemployment following reunification, a young joiner who specialised in renovation restored an abandoned half-timbered house with LEADER support. The building was turned into an art gallery and cultural centre and attracts artists, students and craftworkers. This led on to the creation of the 'ARTigiani' association which works to bring the arts and crafts together. In turn this allowed the small village where the centre is located to become a focus not just of cultural activities but also of training in building techniques using traditional materials.
6. **Dry stone walls in Majorca**. In the Sierra de Tramuntana area of Majorca the disappearance of dry stone walls and farm terraces, due to depopulation, prompted the local council to create a vocational college to train young people in traditional building techniques. In 1991 the local LEADER group launched a support programme to restore the terraces and create a tourist walking trail so that the terraces and other landscape features could be better appreciated. The area is now an internationally renowned skills centre for dry stone building, and many of the area's young people have jobs as a result of the programme.
7. **Old summer houses in Portugal**. In the Vale de Minho area, the LEADER group supported an initiative of the mayor of a small district who wanted to renovate houses traditionally used for summer transhumance and to convert them for tourist use. Thanks to his tenacity and to the fact that this was a strong element of the local identity associated with traditions very much alive in people's memories, negotiations proceeded fast and the project was completed in just two years. One of the houses now serves as a centre for gastronomy and the sale of local products. In parallel, a local visitors' programme was organised around the activities traditionally associated with transhumance, and a tourist company was set up comprising the owners.
8. **Damaged fruit and urban proximity: Rhone-Alpes, France**. About 25 km from Lyon four farmers had long co-operated in marketing their fruit. In 1994 a severe hailstorm hit the orchards, ruining most of the fruit and making it impossible to sell. The farmers invited urban consumers to come to one of the farms, as part of a 'visitors' activity programme, to learn how to make apple tarts using damaged fruit and traditional recipes. The project expanded in scope and 19 local farmers formed an Economic Interest Group to develop the idea further. The result has been the Sunday in the Country project, whereby urban consumers come to prepare their own tarts, patés and apple turnovers. A fruit processing building also serves as a shop for other local products including jam, fruit juices, fruit sorbets and various cheeses and wines.
9. **The Marches Woodland Initiative, England**. Small broad-leaved woodlands are an integral characteristic of the landscape of Shropshire and Herefordshire, and the Marches Woodland Initiative is a five-year

partnership set up in 1997 to increase the value of those woodlands to local communities. At the outset, as much as half of this natural resource was seriously under-utilised and the task was to expand and improve its management both to release economic potential and to enhance environmental value. A consortium of economic and environmental agencies provided capital grants for forestry and woodworking equipment, product development and marketing. Woodland owners received training and advice and jobs were created in woodland management and restoration, local sawmill businesses, cabinet making, the production of other timber products and in tourism, conservation and educational initiatives.

10. **Adding value to women: the UK LEADER experience**. A recent review of UK LEADER II programmes suggested that 'women could be the drivers of rural economic development' (UK LEADER II Network 2000 – section 6) but that many are 'undervalued, lower paid than men, less likely to participate in decision making and inhibited by a range of tangible and intangible constraints'. Addressing those constraints has itself been a significant source of new employment in many LEADER areas, notably in relation to the provision of affordable and accessible childcare and of training programmes, often in outreach centres and often relating to business skills. The review revealed that, thus enabled, many rural women go on to set up and run their own businesses, both on and off the farm.

To conclude, it is interesting to recall the *variety of the resources* to which value has been added in these ten examples; traditional handicraft know-how, a regional language and culture, a specific kind of sandstone, second homes, a redundant historic building, dry stone walls, redundant housing associated with transhumance, damaged fruit coupled with urban proximity, under-utilised woodland and underachieving rural women. Building development strategies on those local resources has everywhere required co-operation between individual entrepreneurs and development agencies and has everywhere delivered benefits that were both economic and socio-cultural.

---

## SEE ALSO...

Case Studies 13, 14, 17, 21 and 22 also portray the local 'adding of value' (but, generally, implicitly).

## SELECTED FURTHER READING

On the theory and practice of the local 'adding of value', the documentation produced by the LEADER Observatory is unparalleled. See the series of five reports (LEADER Observatory 1999a and 2000a, b, c, d.).

# 5

# Entrepreneurship: Backing the Risk-Taker

> If you can make one heap of all your winnings
> And risk it on one turn of pitch and toss,
> And lose, and start again at your beginnings
> And never breathe a word about your loss...
>
> Rudyard Kipling

## DEFINITIONS AND IMPORTANCE

However many audits of potential business opportunities are undertaken by local development agencies, or development strategies written, or elaborate support mechanisms put in place, ultimately someone has to 'make it happen' on the ground. Resources for development may effectively be there waiting – a redundant barn, available labour, a potential market, the prospect of soft loans and pump-priming grants – but a 'spark' is always needed to bring those and other resources together and to turn plans into action. Such is the role of the *entrepreneur*, and in recognition of the crucial part that they play in local development, of the difficulties they typically face and, of course, of the pressing need to diversify many rural economies away from an over-reliance on sectors with limited or declining employment prospects, a good deal of effort by development agencies now goes into trying to create the conditions conducive to the creation and subsequent success of new enterprises.

A *business entrepreneur* has been variously defined as

- someone who sees an opportunity and creates a business as a response to it;
- someone who organises, manages and assumes responsibility for a business; and
- an owner-manager of a firm.

But the key concept is that of risk. Risk-taking is central to entrepreneurship – the risk of losing invested capital, the risk of considerable wasted time, possibly the risk of loss of reputation and social standing. Indeed, following the formula of neoclassical economics, a successful entrepreneur should expect not just a return on his/her labour (a salary) and capital invested (a dividend) but also on the entrepreneurial activity and risk-taking itself (a profit).

A major component of entrepreneurship is the overcoming of obstacles that impede the assembly of the various 'factors of production' so as to make a successful business. Indeed, 'it is not in the perfect market that entrepreneurs seem at their best; surmounting institutional barriers is seen as an entrepreneurial characteristic' (Centre for Rural Economy, 1999: 6). Thus other typical characteristics of business entrepreneurs include (Scottish National Rural Partnership, 1998)

- an eye for opportunities or gaps in the market;
- a desire to achieve – to overcome problems and hurdles;
- a capacity for hard work;
- a readiness for responsibility;
- an ability to take charge and to organise; and
- a readiness to take calculated risks.

Clearly not everyone has those qualities but equally clearly they are essential for any local development programme which is not simply a top-down exercise of delivering the things needed by the people living there – and therefore not a 'development programme' at all. So it is now widely accepted that a key part of rural regeneration is

- the fostering of a culture in which entrepreneurship is valued and flourishes; and
- the delivery of specific assistance to potential and actual entrepreneurs before and during the launch of a new firm and in the difficult period of its establishment and – it is hoped – subsequent growth.

There is another sort of entrepreneur, with many or all of the attributes listed above, who is equally crucial to the local development process. This person is motivated not by the prospect of financial gain but by a wish to create a community benefit of some sort, albeit while reaping less tangible personal rewards such as a sense of achievement and enhanced status or esteem in the community. This is the *community entrepreneur* – a community leader who, in order to make a worthwhile community venture happen, needs the same eye for opportunities and capacity for hard work, problem resolution and risk-taking as does the business entrepreneur. But the risk-taking of community entrepreneurs is of a different sort; they risk not their capital but their reputation in the community and their future credibility. Such people figure prominently in the pages of this book, notably in Chapters 2, 4 and 10.

But while 'community entrepreneurs' seek community benefit rather than personal profit, the sale of goods and services may well feature in their endeavours. If so, they are engaged in *community enterprise*. A 'community enterprise' may be defined as 'any organisation that trades, has a social as well as a commercial purpose, reinvests any surpluses for the good of the community and is democratically owned by its members, its workforce or the local community'[1] – a definition used in the author's evaluation of a regeneration programme in south-west England (Moseley, 2000b: 3) which forms the focus of Case Study 9.

Falling under the community enterprise umbrella, therefore, are such activities as community shops, community pubs, community childcare and transport schemes, credit unions, food co-ops, and many community recycling, composting and renewable energy schemes such as those described in the 'sustainable village' Case Study (number 3) – in effect, any community owned activity which seeks to sell something. Good overviews of community enterprise in a local development context are provided in a New Economics Foundation booklet (undated) which summarises about 30 specific forms that it might take, a DETR guide to good practice (1999), a review of the social economy in the largely rural south-west of England (South West Social Economy Partnership 2000) and, in a more popular format, CRISP, 2001.

Community enterprises are frequently criticised on three grounds. First, that they 'will not create remotely enough jobs in a regeneration area to make an impact on local unemployment in the lifetime of a scheme' (DETR, 1999: 4) – a criticism that may understate the ability of some such enterprises to create jobs and in any event ignores their wider social purpose. Second, that they may be 'inherently fragile or unprofitable, insufficiently commercial in approach and unable to attract enough entrepreneurial talent and commitment' – a common local authority view quoted by DETR (*ibid.*) though then quite properly dismissed on the grounds that much the same could equally be said of many commercial start-ups. Third, that many community enterprises need business support more intensively and for a longer period of time than do most conventional start-ups – a point with some validity but again tending to ignore the wider benefits that often accrue from their operation.

It is important, then, to state that community enterprise normally embraces five objectives:

- generating jobs, especially for the socially excluded;
- providing necessary goods and services which the market is reluctant to provide, at least locally (see the examples listed above);
- enhancing the skills and capacity of a local community;
- empowering individuals and communities to take control of their working lives; and
- recycling income in the community (helping to 'plug the leaky bucket' considered in Chapter 2), notably by reinvesting any trading surplus either in the enterprise itself or in unrelated but worthwhile local community activities.

Reflecting perhaps this wider agenda, the indications are that community enterprise is increasing in importance in rural England (see, for example, South West Social Economy Partnership, 2000; Cumbria Social Economy Forum, undated). And in the context of 'fostering entrepreneurship' it is increasingly seen as well worth investing in – notably for those five valuable social and economic benefits and for the implied spur to broadly based local development.

## SOME KEY ISSUES

Rural locations provide a number of advantages for small businesses and entrepreneurs (North and Smallbone, 1993; LEADER Observatory 1994b; Scottish National Rural Partnership, 1998). Some of these relate directly to the land and the environment, with many new businesses being created in recent years in, for example, tourism and leisure, heritage management, the production of quality foods and the use of woodland for energy and industrial materials. Business surveys have also revealed the high value placed by many rural business people on the environmentally friendly image of their location both as a marketing tool and as an attraction to managerial and other skilled staff recruited from the cities. Also widely commended is the alleged greater flexibility and commitment of the rural workforce and the lower levels of crime when compared with the situation pertaining in many urban environments. In addition, there has been the liberating influence of the improvements in transport and, more particularly, communication technology that have occurred over the past 10 to 20 years and which have eroded, but not obliterated, the penalty of geographical isolation.

Given all of that it is not surprising that micro-businesses – those with fewer than ten employees – accounted in 1999–2000 for 90 per cent of all businesses in rural England (compared with 88.5 per cent elsewhere) and 30 per cent of employment there (UK LEADER II Network, 2000). In addition, the self-employed comprised 12 per cent of the population of working age compared with less than 9 per cent elsewhere (Countryside Agency, 2000a, 2001a).

But set against the advantages of rural locations, the lower population density, relative remoteness and high environmental quality of rural areas also bring problems for the small business person and entrepreneur. Frequently cited in this respect are the problems often experienced in trying to recruit locally sufficient staff with the requisite skills, and the more limited choice of premises available when expansion is envisaged. It is also clear that disadvantages arising from remoteness from markets and suppliers remain despite modern communication technology, and that constraints on business operation and expansion can arise from NIMBY ('not in my back yard') opposition and, more generally, from the high value placed on conserving the rural environment.

The force of these various pros and cons of a rural location for business varies across the country. As a result the challenge for development agencies in promoting entrepreneurship and the creation and success of small businesses is often greater in the more remote rural areas such as Cornwall and west Wales and in those areas with a history of industrial decline and dereliction, such as the former coalfields, than in the more centrally located and generally prosperous counties of England's South East and Midlands.

Also relevant is the culture of those people from whom an entrepreneurial response is hoped. For example, the Cabinet Office Performance and Innovation Unit's review of England's rural economy (1999) noted that while many British farmers have now embraced the challenge of diversification – setting up new businesses on their farms to offset declining incomes from farming itself – many others seem still to be constrained by attitudes forged during a long post-war period in which they were able to see themselves not as 'business people' but as 'producers' operating in a regulated market with guaranteed prices. To have to reappraise their assets and seek to add value to them in enterprising ways while being sensitive to the rigours of a competitive market has proved to be a harsh transition for many farmers and one to which many remain resistant – a cultural problem to be found, of course, right across the European Union (LEADER Observatory, 1994b).

Not, of course, that central and eastern Europe has been immune from such conservatism – quite the contrary. Hansen (1993), for example, has noted that the four decades of communism there were marked by strong top-down decision-making, at best a grudging acceptance of private business and a comforting blanket of security for everyone in terms of their entitlement to some sort of job and the basic income that came from it – a set of circumstances hardly conducive to entrepreneurship. In addition, the lack of experience in those countries of a 'civic society' in which *community* entrepreneurs outside the rigid party system might innovate and take risks for the local community has similarly held back community development and enterprise in the post-communism period.

## THE TOOLKIT

So what can local development agencies do to foster entrepreneurship and the creation and nurturing of new businesses – whether in the 'commercial' or 'community' sector? Not surprisingly, this has been the subject of considerable debate and an outpouring of publications offering advice either to the agencies or to the entrepreneurs themselves (for example LEADER Observatory, 1994b, and undated (a); Ministry of Agriculture, Fisheries and Food 1994a and b; Rural Development Council for Northern Ireland, undated; Rural Development Commission, 1993; Scottish National Rural Partnership, 1998).

Assuming that the basic 'audit' work has been done across the local area – charting the untapped potential of local resources and the area's comparative advantages, and identifying what seem, *prima facie*, to be promising business opportunities (see the chapters on 'diagnosis' and 'adding value'), the task of the local development agency has two components. The first involves trying to improve the 'context for enterprise' with reference to local culture and infrastructure; the second involves delivering specific assistance to actual or potential business people.

### ENHANCING THE CONTEXT FOR ENTERPRISE

'The best way to change a policy environment which hinders entrepreneurship is a participative approach which aims at creating collaborative efforts between government and the private sector to remove constraints and develop strategies of small and medium enterprise development' (Hansen, 1993: 8) – a statement that is in effect an endorsement of the partnership approach to be considered in Chapter 9. Such collaboration and the mutual airing of concerns will often change the perspective both of over-cautious officials – perhaps those with land-use planning responsibilities – and of the more conservative members of business associations and community groups. In short, the experience of local development partnerships is that people do learn and develop by prolonged exposure to the views and concerns of others and by the sharing of responsibility.

Programmes of 'awareness raising' among people who may never have imagined themselves as 'entrepreneurs' can also effect a measure of cultural change – for example programmes targeted at women, young people, small farmers, those recently made redundant and specific local communities where the gap between local opportunities and local entrepreneurship seems particularly wide. Such programmes generally do best if they provide opportunities for those involved to meet and talk to others, like themselves, who have taken the leap and successfully set up their own businesses (a point already made in the chapter on 'innovation'). The work done in recent years among farmers and in coalfield communities by various government agencies provides an example.

Research (for example that of Keeble *et al.*, 1992) has shown that many potential entrepreneurs first migrate from urban to rural areas purely for lifestyle reasons – for reasons unconnected to business considerations – and *then* seek to set up a business. This fact has implications for development agencies as such people need to be courted, and that may mean placing emphasis on environmental improvements, leisure facilities, the availability of appropriate housing and campaigns to persuade talented émigrés to 'come back home'. If just ten out of 100 in-migrants with business ideas and aptitude actually set up a successful business in the area, and if just one of those ten businesses subsequently grows to employ 50 local people, then the attention paid to attracting such people will certainly have been worthwhile.

Local development agencies must also seek to correct deficiencies in the business infrastructure – both 'hard infrastructure' such as roads, sewerage and telecommunications, and 'soft infrastructure' such as education, training and business support services. Some of this gap-filling and upgrading may be undertaken directly by the development agency itself – for example the provision of particular training programmes to improve business and management competence or to improve the basic IT skills of potential entrepreneurs. But more likely the need will be to influence established agencies so that *they* provide the sewerage, serviced industrial land, training courses or whatever is most appropriate for business creation and expansion – an endorsement again of the partnership approach to local development.

Ensuring the provision of appropriate workspace – in the right locations, of the right size and specification and at the right price – is an important component of this. This may involve both the conversion of redundant buildings and new-build. It may include 'starter units' for rent and various kinds of business 'nurseries' or 'incubators' involving the provision on site of some shared services such as secretarial support, childcare and facilities for meetings.

### *THE DELIVERY OF SPECIFIC ASSISTANCE*

The second strand of support for potential or actual entrepreneurs involves targeting individuals with particular needs or problems.

Reaching the right people in the first place and informing them of the assistance available provides an initial challenge and many agencies use a plethora of approaches in the hope that at least one will hit home. Addressing local public meetings may be helpful in generating initial interest in community enterprise; using the mass media or trade press may be appropriate in other contexts, as might circulating publicity literature or working via established business and community networks. Once contact has been made, the most effective way of delivering information, advice and support is often via the deployment of business advisers or project officers of some sort going out to meet potential 'clients' on their home ground. In particularly sparsely populated areas this may involve recruiting part-time, locally recruited representatives, an approach successfully used by a number of Scottish LEADER groups.

Some of the main types of assistance commonly offered by local development agencies to potential entrepreneurs and small business people include:

- information on the various sources of assistance commonly available to new businesses – ranging in the English case from the Small Business Service to the new Learning and Skills Councils; this may involve the development agency in joining in a 'one-stop shop' arrangement, or in signposting to appropriate agencies and organisations;
- access to finance – start-up capital, soft loans etc. The agency may have resources of its own to disburse to selected projects (for example LEADER

groups and various 'challenge' programmes) and/or information and advice regarding other sources;
- securing, and possibly part-financing, some specialist technical assistance needed by an entrepreneur – relating perhaps to business planning or marketing or to particular production processes which might entail feasibility studies;
- help with resolving actual or potential conflicts between the entrepreneur and other stakeholders in local development such as the planning and highway authorities, an environmental agency or a group of local residents; and
- help with 'networking' of different kinds – for example with others in the same line of business, possible suppliers, clients or potential business partners. Help with setting up a village shopkeeper association, a federation of co-operatives or an association of food producers, and arranging 'meet the buyers' events are all examples of such support, which may be particularly valuable in rural areas where individual entrepreneurs may well feel isolated.

## AN APPROACH FOR LEADER 'LOCAL ACTION GROUPS'

As explained in the introductory chapter, LEADER local action groups (LAGs) have sought to promote socio-economic development in their areas largely by offering support to potential entrepreneurs and to small businesses. Conscious that in some cases this was not amounting to much more than seeking out and then part-financing new business ventures meeting certain criteria, the LEADER Observatory (undated(a)) produced advice on 'support systems for new activities in rural areas'. This document enlarged upon many of the tools summarised above and urged the LAGs to adopt a more strategic approach to determining and delivering their support. In particular, it suggested a four-point plan:

- study the needs of entrepreneurs and small businesses in the area and the obstacles they face. In this respect it advised focusing on three types of business: 'fledgling firms' that would normally need help with start-up funding and initial guidance; 'developing firms' which might need support through a period of consolidation or rapid expansion; and 'problem firms' which might need some sort of rescue plan with a restructuring element;
- build relations with other development organisations serving the area since the LAG will not be alone, and other agencies will be responsible for training and specialist business support, for example. The LAG's task is to 'add value', rather than duplicate existing provision;
- decide upon the LAG's own particular role. This may be one of 'signposting' to other agencies plus specific work such as marketing the whole area, providing some specific infrastructure and supplying start-up capital for firms in particular sectors;
- define a communication policy – in relation to the business sector, potential entrepreneurs, other support agencies, business funders and local

elected representatives – so that the LAG's role and capability will be widely understood.

The advice was that, thus prepared, LAGs should then develop protocols setting out how it would actually deal with potential entrepreneurs and with fledgling, developing and problem firms in the sectors of prime interest to its programme. Only then should it start to solicit, select and part-finance new ventures and offer support to business activities already underway.

## COMMUNITY ENTERPRISE: A NEED FOR DISTINCTIVE SUPPORT?

Finally, recalling the earlier discussion of 'community enterprise' as distinct from the more conventionally commercial variety, we must ask whether 'community entrepreneurs' need some sort of special support. The answer seems to be 'yes to some extent', with much of the preceding toolkit being equally applicable to their needs.

Three distinctive requirements are apparent. First is a need for the development agency to target its support as much at voluntary and community organisations with an actual or potential interest in community enterprise as at individual enterprises and entrepreneurs. This 'community development work' has no obvious parallel in the private sector. Second, and more obviously, is a need to give advice and support on matters pertaining exclusively to the community enterprise sector – such as the legal framework of co-operatives, credit unions and development trusts. Third, and already alluded to, is the need to contemplate ongoing and more intensive support for many community enterprises given the breadth of the social and commercial initiative they are undertaking and their frequent, and commendable, involvement of significant numbers of socially excluded people. This should not be exaggerated however; as community enterprise loses some of its 'alternative' image more people with a background in commercial management may well be attracted and the more routine kinds of assistance may in those cases no longer be needed.

All of this may point to a need for support by agencies and project officers specifically selected for their skills at 'the commercial/community interface', as indicated in the case study which follows.

### CASE STUDY 9: COMMUNITY ENTERPRISE IN RURAL SOUTH-WEST ENGLAND

---

'Community enterprise' and the five key objectives that community enterprises typically pursue were set out in the introduction to this chapter. Here we look at some aspects of community enterprise in rural south-west

England, drawing in particular on the author's evaluation of a region-wide Single Regeneration Budget (SRB) programme called CRISP – the Community Regeneration in the South West Partnership (Moseley, 2000b). Its aim was to foster community enterprise by part-funding specialist support staff in the region's county-based Rural Community Councils.

To take one of those RCCs, Dorset Community Action helped significantly over a three-year period in the establishment of 20 community enterprises in that county. These comprised four farming/food projects, three pre-school play or parent support projects, two IT projects, two projects formed to give business advice or to take development initiatives, two learning centres, a furniture recycling project and six other assorted community-based trading ventures. The RCC also gave more modest support to a further 30 local community enterprises including community shops, care groups and trader associations. As for Cornwall RCC, as part of its CRISP programme it provided one-to-one support to a similar number of community enterprises, including 12 potential or 'up-and-running' development trusts. Typically this support comprised advice on constitutional matters, business planning, funding sources and project management. A similar picture pertained in Devon.

The range of community enterprises in the rural South West is therefore a wide one – community retailing, catering, transport, recycling and heritage projects; farmer's markets (where the individual stallholder operates for personal gain but not the communally owned organisations that run them), food co-ops, credit unions, local enterprise trading systems (LETS) and locally focused development trusts which may themselves strive to develop further local community enterprises.

Three specific examples will indicate the variety – but common social purpose – of these community enterprises in the South West.

1. **Development trusts** are community-based organisations working for the sustainable regeneration of their locality through a mixture of economic, social, environmental and cultural initiatives (Development Trusts Association, 1998). They are independent, not-for-profit, locally accountable bodies, often registered charities. Most seek to build up an asset base to give them an income stream with which to sustain their activities in the long term. An excellent example is the Great Torrington and District Community Development Trust in north Devon, one of the fastest growing in Britain. It was formed in 1996 after extensive community consultation following a major loss of employment in the town three years previously. Soon after its launch it successfully bid for £1 million in a national 'Rural Challenge' competition and this and other subsequent funding has enabled it to revitalise the town centre, create jobs and improve various public services.

2. **Somerset Food Links** was set up in 1999 as a response to the crisis in agriculture and the loss of public confidence in food safety. With a mission to promote a thriving local food economy which benefits the whole community it is supported by three local authorities, a health authority and local development agencies. In its first two years it

   - supported several community food projects including food co-ops, community gardens and community cafés;
   - produced directories of local farmers and producers who sell directly to the public;
   - helped set up six farmers' markets in the Mendip and South Somerset districts and a community-based company, Somerset Farmers Markets Ltd, to run them; and
   - supported direct delivery initiatives such as 'vegetable box schemes'.

3. As for the **Bicknoller Post Office/Shop Association,** in Somerset, we may usefully quote from an interview with its chairman, reported in a newsletter of CRISP (1999):

> When we had the shop we were, like many other villages, using it only for a few things, going to the supermarket to save money, and suddenly the shop was gone. People started saying 'why can't we have the shop back?' and we were lucky enough to hear about VIRSA (the Village Retail Services Association). They explained how to become a Friendly Society and all the ins and outs of running a shop. After a lot of door knocking we raised nearly £6000 and the Rural Development Commission promised another £2000. We were all ready to start converting a stable we had been offered but that fell through, which was a real blow. We spent ages looking at other options and in the end the village hall committee agreed to let us convert an old barn on their site. We had to ask for funds again and raised another £5000 from shares.
>
> A lot of people said it wouldn't work but the key to it was when we decided to have two staff on duty all the time – it meant the people working there felt secure and more confident. We now have 30 volunteers helping and over half the households in the village are shareholders which is pretty fantastic.
>
> You need to be very clear, though, that this isn't the sort of business that could give someone a livelihood. All the staff are volunteers, though we are able to give the shop manageress a small honorarium. The sub-postmaster contributes the £1600 a year fee to the Association. Basically we only make ends meet because there's such strong support in the village.
>
> After the third year, a lot of the shareholders were so pleased they said they didn't want their money back and that happened again the next year so we've now paid everything off and are building up a surplus. The post office opens four mornings a week and the shop six mornings and two afternoons selling more or less exactly what our customers ask for – but no newspapers (too much trouble) and no cigarettes or alcohol for security reasons ... It's changed the village beyond belief.

Those three examples demonstrate clearly that entrepreneurship is not the exclusive preserve of the private sector, but they also reveal the community entrepreneur's need for support – support that typically has two elements: first, general 'development work' – raising awareness, enhancing the 'capacity' of communities and key individuals, building up networks to facilitate the exchange of experience and contacts; and second, 'case-work' – giving expert one-to-one support to community organisations planning the launch of a community enterprise and to actual enterprises that have taken the plunge.

---

## CASE STUDY 10: 'SCOOPE': SCHOOL CHILDREN ORGANISING AND OPERATING PROFITABLE ENTERPRISES IN TIPPERARY

---

Conscious of a need to promote positive attitudes to enterprise at the earliest opportunity in a part of rural Ireland where an entrepreneurial response to the area's social and economic problems had been often lacking, the Tipperary LEADER II group came up with an innovative response (LEADER Observatory 1999g). It devoted some 15,000 euros to an initiative to inculcate an enterprise culture and develop enterprise skills in the area's children. Interestingly, a 'generation of ideas' experiment in the area's schools found that most young people aged over 13 years seemed already locked into career paths that had little to do with the development of enterprise, while those aged 5 to 12 produced much more enterprising ideas; so the project concentrated on primary schools.

It was structured to align with the arts and crafts, mathematics, English and communication elements of the curriculum. A project pack and publicity posters were produced dealing with market research, production, marketing, accounting etc. Thirty schools were visited and a pilot group of nine selected to embark on the project in autumn 1997. Each school was allocated 60 euros as a start-up fund with the opportunity to apply for more when needed as well as the possibility of winning a cash prize at a project presentation day to be held a year later.

The children were encouraged to develop enterprising ventures themselves, with the teachers and the LEADER group's development officer acting as facilitators. Pupils organised themselves into management, production, financial and marketing teams and kept records of the difficulties encountered and the changes of plan which were needed as their projects progressed.

There was great variety in the enterprises established. They included the publication of a heritage booklet, production of a heritage video interpreting local artefacts, production of Christmas cards, gardening and the sale of flowers and vegetables, crafts for a school fair, wood products and crafts, services such as coffee and crepe mornings, potted plants and decorative containers, paper collection for recycling, paper crafts and sweet-making, production of a video and newsletter on the theme of 'life in an Irish school', and publication of a book on parish activities.

The profits made by the individual schools ranged from about 45 to 1000 euros, and these were ploughed back into school activities and facilities, but the emphasis was not on the profit generated but on the learning process and the development of enterprise skills.

Over 1000 people attended the exhibition day in 1998 at which each school presented its project and received commendation. The whole exercise was judged a great success especially as it had managed to involve children with different skills and aptitudes, all of whom learned something about self-reliance and enterprise and about working together towards a common goal.

Other benefits included capturing the interest of local parents, neighbours and sponsors, with much enhanced dialogue between the generations as a significant by-product of the project. Moreover, the various books and videos that the children produced, plus the media coverage of the whole project, served to raise the profile of the schools, the area and the need for a 'can do' response to the deep-seated socio-economic problems of the area.

SCOOPE continued for two further years until the money ran out in 2000. By then the number of schools participating had increased to 14 and local development specialists from other LEADER groups in Italy, Spain, Scotland and Wales had visited to see if the idea might be transferred to their areas. It was certainly an innovative approach to the fostering of entrepreneurship, and as a member of the Tipperary LEADER II group remarked to the author in 2001, 'it's too early to tell but we think we've sown the idea of self-employment in the minds of some of the youngsters at least.'

---

### *NOTE*

1. Here we will consider the terms 'community enterprise(s)' and 'social enterprise(s)' as effectively synonymous, though only the former seems always to have a local remit. The 'social *economy*', however, is broader; it also includes grant-funded organisations such as charities, community groups that do not trade and other voluntary organisations.

### *SEE ALSO...*

Several other case studies relate in some way to entrepreneurship and its support – for example 7, 16, 26 and 27.

## *SELECTED FURTHER READING*

On entrepreneurship, see Hansen (1993); LEADER Observatory (undated (a)); and Scottish National Rural Partnership (1998). On community enterprise, see New Economics Foundation (undated); DETR (1999); and CRISP (2001), the latter making reference to many useful websites.

# 6

# Community: Promoting a Sense of Belonging

> 'No man is an Island, entire of it self; every man is a piece of the Continent, a part of the main... And therefore never send to know for whom the bell tolls; it tolls for thee.'
>
> John Donne

## DEFINITIONS AND IMPORTANCE

The promotion of 'community' has become a major feature of rural development programmes, indeed of social policy more generally. This is for three reasons; first, 'community' is valued as a 'consumer good' with a lot of people longing to be in what they perceive to be a real community; second, it is valued as a resource, a lot of agencies striving to enhance and to exploit it; and third, it is for various reasons also at risk. But, unfortunately, it is an elusive concept – 'community' is as difficult to define as 'communities' are difficult to identify and to promote.

As Butcher put it, 'community is one of those "hoorah" words that seem to encourage warm and positive feelings at the expense of precise and meaningful analysis' (1993: 3). And certainly this sense of community being self-evidently a 'good thing' goes deep; to quote Rogers (1993: 2), 'the rural community is seen as the essence of the English good life – a collection of people well integrated into their local society and their environment and living productive and rewarding lives'.

The problem is that the concept is just so elusive that no-one can claim that their definition is the right one and everyone else's is wrong. In a recent comprehensive review, Liepins (2000) stressed that 'community' is a socially constructed and therefore contestable concept and that, in consequence, communities cannot simply be presumed to exist in some hidden form while awaiting the more rigorous definition and better research tools which would reveal them unambiguously in the real world. But we must nevertheless try to 'unpick' the term as it is frequently used and to this end it is useful to consider the word first as a noun and then as an adjective.

### THE NOUN 'COMMUNITY'

All commentators agree that the term 'community' cannot simply be applied to any collection of people who happen to live close to one another. Rather, people in communities must share something in common and, through sharing that something and being conscious that they do so, interact. These notions of 'consciously sharing' and of 'interaction' are at the heart of most definitions.

Going on from that, it is important to distinguish between *interest communities* where physical proximity is not a requirement – for example the Jewish community, the gay community or the academic community – and *territorial communities* where it is, for example, the residents of a village or urban housing estate who identify strongly with that place. A territorial community, of course, can be particularly strong when reinforced by an interest community, as with the close-knit mining communities of the former English and Welsh coalfields and the Catholic community of Derry's Bogside – each of those communities, incidentally, further cemented by a third shared characteristic, namely adversity and threat. Note that the expression 'close-knit' graphically suggests the interweaving of two or more mutually supportive threads.

With such considerations in mind we will adopt here the definition of a (territorial) community put forward by Rogers, based on earlier work by Newby and others: 'a community is a collection of people in a given locality who share a common identity and who interact to form some sort of local social system' (1993: 5). Of course, that definition makes no reference to scale – but the convention seems to be that neither the individual household, however extended, nor whole countries or regions, such as Wales or the north-east of England, qualify. In short, 'territorial communities' exist at some ill-defined intermediate geographical scale and it is the enhancement of such communities that figures so prominently in rural and local development programmes.

### THE ADJECTIVE 'COMMUNITY'

Put before a noun, the word 'community' becomes even more elusive – and not just because the resulting terms have often been coined, at least in part, for the somewhat cynical political reason of capitalising on the 'warm and positive feelings' widely associated with it. The difficulty arises also from what may be called the 'preposition problem'; does the term imply something *for*, *by*, *in*, *with* or *of* the community? In fact this varies, as the following list of commonly used terms makes clear: community nursing (in?); community transport (by? for?); community arts (by?); community strategies (for?); community development (of? with?); community enterprise (by? of?); community policing (in?); community charge, alias the 'poll tax' (of?); community care (for? of? in? by?).

The point is that these and other acts of service delivery and public policy imply different interpretations of, and relationships with, 'the community', usually but not always reflecting a wish to enlist the latter as an ally if not a resource. But implied also is a sense that the policy is, or is designed to appear to be, based on certain 'community values' – of solidarity, involvement, consensus and sharing.

However, for the student or practitioner of rural development, the important point is that, despite being an elusive, exploited and somewhat fuzzy concept, community *matters*. It matters both to the resident and would-be resident of rural areas and to the agencies and organisations that serve them. Increasingly, weight is placed on deliberately fostering 'community' as an act of policy, and the next sections consider why that is so and how it might be achieved.

## SOME KEY ISSUES

We are taking 'territorial communities', then, to comprise collections of people who, at some relatively local geographical scale,

- live in reasonable proximity;
- feel they belong to a particular place;
- share a common identity; and
- interact with one another.

Note that this interpretation does not imply that they need actually to like each other, or even that they will agree on anything other than 'where they are coming from' in both the geographical and popular uses of that expression. Both diversity within communities and some measure of conflict seem to be the norm.

Diversity exists in that communities can and do exist within communities, or, as Russell put it, 'the modern village is a community of communities and village life is the product of the interaction between them' (cited in Rogers, 1993: 5). Indeed, following Newby (1985), we may note 'encapsulated communities' in many contemporary English villages, for example a rump of farmers and farmworkers surrounded by newcomers with very different values and attitudes. And conflict is hardly surprising, given that the interests of farmers and conservationists, commuters and young families, retired people and teenagers, the middle-aged 'NIMBYs' and advocates of development are likely to diverge significantly, even if beneath the conflict there remains some measure of common identity born of attachment to place and mutual, if sometimes strained, interaction.

Two other frequently claimed or assumed elements of 'territorial community' warrant consideration. First, does the local social interaction need to be *caring*, rather than, say, wholly businesslike or political? Even if this does not figure in our definition, two strands of evidence suggest that this caring dimension is in fact to be found in a good deal of the social interaction occurring within local communities. First is the plethora of 'good-neighbourly' organisations that exist in English villages, from young farmers' clubs to village care groups. Second is the apparent greater readiness of people to support good causes if they are overtly aimed at *local* people. Risking a sweeping generalisation, without hard evidence to support it, the norm of conflict in England's rural communities does not, it seems, override a measure of enhanced concern for one's neighbours in need, as distinct from such people nationwide. 'More affordable housing for low-income locals' may just stand a chance of acceptance; 'more affordable housing for low-income people coming out from Birmingham' has no chance at all. Of course, a realisation that this local identification exists underlies much of the 'community dimension' of public policy referred to earlier and again below.

Second, what of *social balance* which we may define as an approximate replication at the local level of the demographic, social, cultural and ethnic proportions found nationally. Can a collection of local people, interacting and sharing a strong sense of place, still be a 'community' if it is very largely comprised of just elderly people or just middle-class people or just young families? The answer must be 'Why not?' Indeed, in that an 'interest community' is thereby more likely to reinforce the 'territorial community', then the latter may well be stronger for being relatively homogeneous.

But, in the context of any policy to promote *sustainable* rural communities, this may be short-sighted. Balanced communities seem likely to be more sustainable in the long term, with each locality able to 'internalise' more of its economic and caring activities (as with poor women at home having wealthy households nearby from whom they can earn a little extra money, and frail elderly people having young people close at hand to care for them, for example). In addition, sustainability may be enhanced by the balanced community's erosion of 'polarisation', meaning ignorance of and indifference towards people who are 'not like us' – ignorance and indifference that would threaten any sense of 'community' on a wider, national canvas.

So why is 'community' valued? Really this subsumes two questions. First, why do *residents* value it? This surely stems from some of its key attributes already discussed. Communities are, or are seen to be, at a more 'human scale' than are the large, allegedly amorphous, urban areas that many people are used to. They are perceived to involve knowing one's neighbour, greater friendliness, the prospect of mutual support and greater community spirit (a perception confirmed in several social surveys summarised in Countryside Agency (2000b)). They suggest a shared identity and commitment to a place. If it were possible to rank the reasons why such a premium is placed on living in village and small town locations by those who can afford to choose, it seems likely that 'being in a real community' would emerge second only to environmental quality – even if the massive influx of affluent people into England's villages has, ironically, served to erode more than enhance the very 'community' that many have sought to buy into.

The second question is 'Why do *public sector agencies* value territorial communities?' The main answer is surely that 'community' is a resource that can be tapped into. Local communities are seen to embody valuable attributes – a culture of mutual support, a measure of altruism, informal networks, local information, energy and vitality, a capacity to reach isolated people – which the cost-conscious state would be foolish to ignore; hence, many of those community-focused services referred to earlier. But this is not the only reason; increasingly, fostering local communities is seen as a contribution to the goal of sustainability, with their short-distance support networks and economic transactions reducing the need to travel to or from major urban centres. The fully 'self-sufficient local community' is surely a pipe-dream, since some things – such as brain surgery – simply cannot be provided in every locality, but this does not negate the value of encouraging steps in that direction.

For all of these reasons, there has been growing concern about the powerful forces that combine to *erode territorial communities* in rural areas. Among these, and closely intertwined, are:

- the mobility afforded by two-car ownership which makes it easy to commute long distances and to enjoy social and cultural activities way beyond one's place of residence;
- the demise of many village services such as shops, schools and pubs and, thereby, of meeting-places where ideas, concerns and gossip can be exchanged;
- demographic change, including the loss of young people with their capacity to reproduce and invigorate community, and the disproportionate in-migration of people sufficiently wealthy and well-connected not to need community as an essential feature of their lives, even if the idea of 'community' was part of the initial lure; and
- increasingly individualistic lifestyles, including leisure pursued at either the 'intra-household scale' (watching television, drinking at home, surfing the Net, gardening) or the 'extra-village scale' (visiting relatives and friends some distance away), each scale ignoring the immediate locality.

And, of course, if we were to accept that 'social balance' is indeed a contributor to 'community', then other 'eroding forces' come into consideration. Most notable – with each of them tending in essence to favour a build-up of the 'middle classes' at the expense of those on lower incomes – are the transference of council housing onto the open market, and restrictive land-use planning policies which necessarily push up the price of development land and thereby favour the construction of more expensive new homes.

As for the *aspiration to community*, we must of course beware of what Newby termed 'rural retrospective regret' – of hankering back to, and seeking to recreate, some perceived idyll of former years. Not only did that idyll rarely exist, except in romantic literature, but its replication, even if it were possible, would certainly not meet the needs of a modern society and economy. That said, there does seem to be great value in trying to protect where it exists, and to promote where it does not, the qualities of 'community' as described in this chapter.

A catchy way of summarising this aspiration is to refer to the A, B, C, D and E of the attributes of communities to which many people and policy-makers aspire. Village communities should be:

- Active – socially, culturally and economically;
- Balanced – meaning broadly mirroring the nation's demographic, socio-economic and ethnic mix;
- Caring – with a culture and local institutions that foster a concern for the well-being of one's neighbours;
- Durable – meaning sustainable with regard to the use of environmental, manufactured, human and social capital (concepts explained in Chapter 2); and
- Empowered – with appropriate decision-making devolved from higher levels.

That may seem idealistic, but moving significantly in that direction is not impossible and below we consider the relevant toolkit.

## THE TOOLKIT

There is now considerable experience of trying explicitly to promote 'community' at the local level in Britain's rural areas. What we are concerned with here are any initiatives that might promote the ABCDE model set out above – that might foster territorial communities at the village, parish or small-town level that are active, balanced, caring, durable and empowered.

Arguably, the single most powerful tool in this respect is, effectively, unacceptable as a deliberate act of policy. It is to *threaten some perceived misfortune* which would bring the community together to oppose it. In short, nothing is more effective in generating community spirit than the prospect of a bypass through a treasured landscape, a large speculative housing estate, a new mineral-working or a school closure – though the camaraderie thus engendered will often last no longer than the ensuing campaign, whether it is successful or not.

Turning to more practical propositions, we can be brief as the case studies that follow – and some in other chapters – provide greater detail and concrete examples.

Without implying 'physical determinism' (the idea that the physical environment determines human behaviour) it is still true that *land-use planning* has a significant role to play in encouraging 'community'. First is its ability to affect the *pace and scale of new development* in the various villages and small towns of a local authority area. The suggestion is that neither an embargo on any new development nor change so rapid and substantial that it 'swamps' whatever community already exists is likely to be helpful. Modest, steady development, allowing for an adequate provision of appropriate new housing and employment, seems more likely to foster the social mix, influx of new households and local economic activity that thriving communities need. Second, by shaping the emerging *land-use pattern and the routeways* that connect the different parts of the village or town, social interaction can be enhanced. Socially mixed housing developments and a layout that encourages walking and cycling between and within the different zones of the settlement in question are each more likely to favour social interaction than would a strict demarcation between housing estates and a land-use pattern that virtually necessitates car use.

The land-use planner can also help in other ways – for example by encouraging genuine community involvement in the planning process perhaps through the preparation of 'village design statements' by local communities, and by resisting change-of-use applications in relation to village service outlets such as shops and pubs, in the absence of convincing evidence that their continued operation would be commercially unviable and socially irrelevant.

*Housing policy*, coupled with land-use planning policy, can have a tremendous influence upon the viability of 'community', as it holds the key to social balance. The need is to ensure sufficient supply and retention of affordable housing for *local* and for *low-income* households – two groups which are not synonymous but which will certainly overlap. 'Homes for locals' are important in community terms since they help the local retention of extended families and foster greater longevity of occupation and hence of 'attachment to place'; and homes for the relatively poor generally help a village's social mix, the importance of which has already been stressed.

But how to increase the supply of affordable housing earmarked for local, low-income households? In essence this means ensuring that sites are available at viable prices, defusing NIMBY opposition and making more funding available to housing associations and other 'registered social landlords'. See Chapter 7, and the Rural White Paper (Department of Environment, Transport and the Regions and Ministry of Agriculture, Fisheries and Food, 2000) for a more detailed consideration of this vexed issue.

*Retaining meeting-places* such as shops, post offices, schools, pubs and clubs, churches and village halls/community centres is important when so many have been lost and others are under threat. As well as providing venues for social and cultural activities they also make it easier for people to check on the well-being of the vulnerable, to exchange information and ideas and to plan various local activities. Again, there is much that can be done to retain them, including offering business rate relief which has been available to selected local service outlets since 1998, assisting with capital expenditure on the refurbishment or upgrading of buildings and equipment, giving business advice and information to 'outlet managers' such as shopkeepers and to groups of local activists contemplating taking over a faltering shop or pub, and steering more business into village outlets, for example banking services in post offices and adult education in schools. Attempts to promote the joint provision of disparate services, for example post offices in pubs or in filling-stations, can make sense especially if this helps to sustain a village's last service outlet.

But undoubtedly the most powerful way of building active, caring communities is deliberately to *empower* very local organisations to carry out activities that have hitherto been the province of 'higher up' agencies. Of course, this poses problems. The readiness of people at the very local level to accept new responsibilities will vary greatly from place to place with the result that 'peaks and troughs' of service provision could easily replace the 'gently undulating plateau' that comes from larger agencies delivering services in accordance with national standards and some sort of strategic plan. And, of course, there is the problem of a lack of economies of scale – a major reason why we in Britain have resisted the continental model of delivering quite sophisticated services, such as school building maintenance and some policing, at the 'commune level' and why the continental country with the most 'micro-local authorities' (France, with its 32,000 communes having fewer than 2000 inhabitants) is increasingly creating 'inter-commune' partnerships to do what is patently uneconomic at the individual commune level.

Nevertheless, as underscored in the Rural White Paper (DETR and MAFF, 2000), there is considerable merit in

- permitting and encouraging those parish and town councils that are both ready and willing to deliver some services on behalf of the county and district councils;
- helping parish councils working with local groups to develop and implement their own wide-ranging 'community action plans' – as exemplified in Case Study 24;
- helping local communities to launch and manage a variety of co-operative ventures such as community shops, community transport and pre-school and after-school care;
- supporting other local agencies and initiatives such as development trusts, good neighbour schemes, local amenity groups and adult education provision; and
- more generally, fostering a culture of local management and enterprise – for example in schools, village halls and community resource centres.

All involve 'subsidiarity' – pushing power and responsibility down to the lowest feasible level – and developing communities through deliberate empowerment. But none of this can be achieved without putting in place sophisticated machinery to offer training, technical advice and other forms of support and without earmarking sufficient money for project grants and the building of 'local capacity'. Government programmes, such as Rural Action (see Case Study 28), which underwrote thousands of local environmental initiatives in the mid-1990s, LEADER and the new Vital Villages initiative (Countryside Agency, 2001b) exist to do just that. But crucially important are 'intermediate voluntary bodies' such as England's Rural Community Councils and Councils for Voluntary Service, which operate at the district or county level to support local ventures, mainly through deploying fieldworkers to work alongside local activists and to help make their initiatives a success.

All in all, it would be wrong to conclude that the battle to retain and even enhance active, balanced, caring, durable and empowered rural communities is being lost. There are deep-seated social, economic and cultural forces rendering this battle a difficult one. But thousands of small successes each and every year give cause for guarded optimism.

## CASE STUDY 11: THE PROMOTION OF 'COMMUNITY' BY ENGLAND'S RURAL COMMUNITY COUNCILS

In England there are 38 county-scale charitable organisations called Rural Community Councils, or RCCs, which, collectively, cover the whole country. They work to support the local voluntary sector, to improve the quality of life of rural people, especially the more disadvantaged, and to promote 'community development' at the village or parish scale. 'Community development' is commonly construed as the process of facilitating the active participation of local people in collective efforts to improve their quality of life and that of their neighbours, but here we will interpret the term a little more narrowly and explore how the RCCs actually promote 'territorial communities' in the sense in which that term has been used above.

Though each serving perhaps 250 or 300 parishes across their counties, RCCs typically have a staff of only around 20, though these are often supplemented by a variety of volunteers who can provide specific assistance. RCC funding comes from a plethora of sources including central and local government, the sale of services to various organisations, charitable donations and business sponsorship. The essence of their work, indeed of all community development , is to 'make a little go a long way' by working in partnerships and by doing just enough to galvanise others into constructive action.

Here we will consider the work of just five RCCs – those serving Cumbria, Gloucestershire, Herefordshire and Worcestershire, Oxfordshire, and Shropshire – and, drawing upon their recent Annual Reports, give examples of their efforts to foster five particular elements of 'community' which were considered above. These are:

- the increased local interaction of people;
- a shared sense of belonging to the local place;
- increased mutual care at the local level;
- maintained or enhanced social balance; and
- empowerment – placing more power in local hands.

**The increased local interaction of people**

*Cumbria*
Various initiatives to retain or improve village service 'outlets' in the smaller communities, especially through their multiple use, for example a post office opened in a village hall and a library in a community centre.

*Gloucestershire*
Deploying a community development worker at a former RAF base now sold off for housing and, through her, helping to launch a community association to bring the new residents together.

*Herefordshire and Worcestershire*
Developing a network of small town 'community resource centres' which provide a range of facilities, such as IT and printing equipment, in support of self-help by local people.

*Oxfordshire*
Running training courses for volunteers who produce some of the county's 180 village newsletters.

*Shropshire*
Working with other agencies and local people to get new village halls built and older ones refurbished.

**A shared sense of belonging to the local place**

*Cumbria*
Helping a community association to fundraise for, and eventually manage, a new community centre in a village suffering from steady social and economic decline.

*Gloucestershire*
Organising a participatory 'planning for real' exercise so that local people might better articulate their ideas for improvements to their town.

*Herefordshire and Worcestershire*
Helping a number of communities to produce their own Village Design Statements to advise the planning authority on the preferred physical development of their village.

*Oxfordshire*
Managing a 'Best Kept Village' competition which encourages residents to care for their village, especially its public spaces.

*Shropshire*
Supporting 'village photography' and 'parish map' projects whereby local people record what is distinctive about their village and then display the photographs or maps for local people to see and discuss.

**Increased mutual care at the local level**

*Cumbria*
Helping to launch the 'Muncaster MicroBus' – a community-run vehicle adapted for wheelchair-users and for parents with young children and heavy shopping.

*Gloucestershire*
Launching a mobile day centre for elderly people, serving several villages and bringing together local volunteers and voluntary organisations.

*Herefordshire and Worcestershire*
Arranging training for playworkers – mainly women keen to improve play and childcare provision in their locality.

*Oxfordshire*
Supporting 240 'parish transport representatives' – volunteers who help to identify local transport needs and encourage a community response where appropriate.

*Shropshire*
Managing a county Village of the Year competition in which participating villages are assessed on several criteria including the quality of informal care for older people and the vitality of village life. (Case Study 12, presents a *national* winner of this competition.)

**Maintained or enhanced 'social balance'**

*Cumbria*
Helping to create an up-to-date inventory of resources for young people in rural Cumbria and supporting the 'diversification' and viability of small schools – two ways of helping families with young children to continue living in small villages.

*Gloucestershire*
Managing 'Jump-Start' in several villages – a scheme whereby young people are helped to live at home while borrowing the motorised transport necessary to get to work elsewhere.

*Herefordshire and Worcestershire*
Working with a local housing association to create a Rural Housing Enabler post in one rural district and thereby an improved ability to deliver affordable housing in villages where it is needed.

*Oxfordshire*
Helping several small parishes in the county to lobby the government to be included on a list of parishes exempt from the tenants' right-to-buy scheme, thereby preserving the local stock of affordable housing for rent and the ability of low-income families to live in the area.

**Empowerment – placing more power in local hands**

*Cumbria*
Supporting local 'community transport groups' in their operation of community minibus services.

*Gloucestershire*
Advising several communities with recently completed 'village appraisals' on how they might turn their hopes and plans into action.

*Herefordshire and Worcestershire*
Supporting a partnership of voluntary, community, statutory and business organisations in the Tenbury area in their successful bids for 'Rural Challenge' and other funding, thereby launching a powerful partnership serving the local community.

*Oxfordshire*
Providing training and advice for parish clerks and councillors and thereby helping parish and town councils to be more effective in serving their local communities.

*Shropshire*
Managing the county's Rural Action initiative which, in any one year, channelled funding to about 60 local groups intent on improving their local environment. (See Case Study 28 for more detail on Rural Action)

## CASE STUDY 12: THE NATIONAL 'VILLAGE OF THE YEAR' 2000

Our second case study is of a national competition designed to identify, celebrate and, in a modest way, reward truly outstanding village communities. The annual England and Wales Village of the Year Competition was launched in 1997 having evolved from the countywide Best Kept Village competitions which, for 50 years or more, fostered community pride in the built environment and physical appearance of our villages.

Though about 18 Best Kept Village county competitions still remain (as compared with 26 Village of the Year competitions) their formula seems a little dated in today's climate, with their emphasis often tending, in practice, to be focused more on 'keeping things neat and tidy' than on truly promoting the vibrancy and mutual caring aspects of village life. Certainly, a competition emphasising care for a village's people rather than for its visual appearance seems better attuned to the holistic perspective of 'community development' espoused in this book.

Village of the Year is sponsored by Calor Gas and the *Daily Telegraph*, each sponsor keen to promote its profile in rural communities. It comprises a series of interlocking county, regional and national competitions in which England's Rural Community Councils, county branches of the Council for the Protection of Rural England and other voluntary organisations play a facilitating role. The judges at all three levels are people with expertise and experience of village and community life.

Two aspects of the competition are of interest in the context of this chapter on the nature of 'community': What are the criteria used for selecting 'villages of the year'?, and What was so special about the national winner in 2000 – Great Bentley in Essex?

As for the selection procedure, all villages entering the competition are judged according to five broad criteria which reflect some measure of consensus between the various agencies involved. (A sixth criterion, 'the provision and use of IT by the community', was added in 2001.) The five criteria are fleshed out in a series of pointers to guide both the local communities in producing their written submissions and the various judges who read those submissions prior to visiting the villages and questioning a number of people living there.

The five criteria, with some of the associated pointers and questions, as gleaned from the guidance issued at county and national level, are:

*Care for the village environment*
For example, Do you have an active conservation group? Who is involved? What does it do? Do you have a recycling scheme? How are environmental issues and activities communicated in the village? Is there a 'village policy' or plan for the environment? Are the village school and/or other organisations actively involved? What practical schemes to improve the village environment have been planned or recently completed?

*Business and the village community*
For example, How does the local community support local business and vice versa? What employment opportunities do local businesses offer to local people? Are there any community-run enterprises? Is local produce sold in the village? Has the community had to deal with the loss of a local business such as a shop or post office, and if so, how? What is the parish council's attitude to local business?

*Young people*
For example, What services, activities and facilities for children and young people are available in the village, such as crèches, pre-school care and youth clubs? How are these supported by the community? What events are organised especially for young people? Are young people encouraged to help make the decisions that affect their lives? What do young people contribute to the local community? Are young people helped by the community to gain access to facilities outside the village?

*Older people*
For example, What services, activities and facilities for older people are available in the village – such as car schemes and social clubs? Is there a village care scheme? Are there community-run initiatives for older people, for example regular outings, luncheon clubs and help with shopping? Is there a housing scheme specifically to help older people stay in their own community? Are older people involved in volunteering, for example reading in schools, or participating in oral history groups? Are older people encouraged to get involved in local decision-making?

*Community life*
For example, Give a concise list of the activities that go on in the village, as if for a newcomer. How does information circulate in your village? Are there any village charities – if so, how are they publicised and who benefits from them? Does your parish council consult and communicate well? Give examples of 'village problems' (such as safety and security) and how they have been dealt with by the community. Is there any collective forward planning in your community? Is the village hall genuinely accessible to all and how democratically is it managed? How are newcomers welcomed into the village?

These are tough criteria. Certainly the smaller villages could hardly be expected to score well on more than a handful, so it is appropriate that at the county level there are prizes for villages in different population bands, and also for those that do outstandingly well on just one of each of the five criteria.

The national winner in 2000 was the village of Great Bentley, near Colchester, with a population of some 2200. Its village appraisal report, published in November 1999, revealed it to have a dozen shops including a farm shop; two pubs; a surgery accommodating six doctors and other health professionals; a dispensing pharmacy; a school with 175 pupils and an active parent–teacher association; a large, well-used village hall; Anglican and Catholic churches; a Methodist chapel; and a 43-acre village green, much used for carnivals, shows and a range of sporting and cultural events. The village also has a small industrial estate which was developed in the 1960s to provide jobs for locals and now accommodates 20 small companies employing some 250 people. Such an impressive range of facilities is certainly a resource for the community and, in some measure, also the product of the community's endeavour.

But what is the evidence for Great Bentley being a truly vibrant and caring community? Here we can mention, in relation to each of the five criteria, just a few things that impressed the judges:

- **Care for the environment** The large village green is managed by the parish council with the active support of the 'Friends of the Green'. A parish wildlife survey has been carried out drawing on information from local people. An environmental clean-up day was a resounding success and is to be an annual event.
- **Business and the community** This is a *working* village; of the 250 people working on the industrial estate, 60 per cent live in the parish, and the other 31 businesses in the parish employ a further 150 people. In all, about 40 Great Bentley businesses take advertising space in the monthly *Parish News*.
- **Young people** There are four pre-school groups for local children. The Great Bentley scout hut is used for a variety of purposes by children from 120 families in the area. In 2000 it was undergoing extensive repairs and refurbishment, carried out by local businesses and largely funded by local initiatives, many of them organised by the children themselves.
- **Older people** The Great Bentley Good Neighbours scheme provides a comprehensive voluntary community care service, with 62 'road stewards' ensuring that people's needs are identified and with help available across the whole parish. There is a 'medical aids' loan service funded and run by local volunteers.
- **Community life** Fifty people are involved in producing the *Great Bentley Parish News* each month, which has twice been judged the best

village magazine in Essex. Its 'What's On?' diary in the 60-page October 2000 issue, for example, listed 24 events or activities taking place in the village that month.

By any standards Great Bentley scores highly on all the 'indicators of community' suggested earlier in this chapter – solidarity and caring; a shared sense of place; a common identity; local social interaction; widespread involvement and social balance. It is truly a 'territorial community'.

---

### SEE ALSO ...

Case Studies 3,19, 20 and 24 all portray some aspect of 'community' or its promotion.

### SELECTED FURTHER READING

Liepins (2000) reviews the theory of 'community'; Butcher *et al.* (see Butcher, 1993 in the references) and Rogers (1993) link theory and practice, and Derounian (1998) is wholly concerned with practice.

# 7

# Social Inclusion: Bringing on Board

My mother said that I never should
Play with the gypsies in the wood.

Nursery rhyme

## DEFINITIONS AND IMPORTANCE

On many indicators of well-being or prosperity rural Britain performs significantly better than does the nation as a whole. But such favourable statistics, relating, for example, to levels of unemployment, to rates of job growth and of small-business creation, to average incomes and indicators of good health, conceal considerable variations from area to area and between households living in the same area. And there are other measures, for example, relating to the affordability of housing or to the ease of access to employment or to training, which tell a different story. (See the 'State of the Countryside 2001', report produced by the Countryside Agency (2001a).) At the level of the individual, the space and environment that characterise rural areas can be a joy to those with a decent home, a well-paid job (or adequate pension) and their own transport – but a prison for those without these things.

This chapter explores the social and economic deprivation to be found within rural areas, but from a particular perspective. It will focus directly not on *disadvantage*, a term now normally taken to mean 'the inability of individuals or households to share in styles of life open to the majority', nor on *poverty*, which relates to the portion of that inability which is due to not having enough money. Rather, it is concerned with *social exclusion* – a term little used in Britain before the mid-1990s but now widely employed to mean 'a multi-dimensional, dynamic process which refers to the breakdown or malfunction of the major systems in society that should guarantee the social integration of individuals or households...it implies a focus less on victims and more upon systems failure, especially on the processes which cause exclusion' (after Shucksmith, 2000). On 'process', see also Policy Studies Institute (1998) and, for more human accounts, NCH (undated) and, Countryside Agency (2000a)).

Such an approach, then, shifts the focus from the individual to his/her context, from a primary concern with income to one which is multi-dimensional, from a snapshot of a situation to an analysis of process, and from the 'arithmetic of woe' to the evidence of systems failure. It is an approach that is better able to inform policy and practice since, at its most simple level, it argues 'establish the dynamics of exclusion and you have the basis of action for inclusion'. We will therefore look both at the processes of exclusion in rural Britain and at the 'toolkit' which is available for reversing, or at least mitigating, those processes. We will then go on to consider the potential suitability of area-based rural development programmes to deliver at least some of the desired 'inclusion'.

The analysis and attempted alleviation of social exclusion in rural areas is afforded considerable importance today for two reasons:

- it produces *social injustice* on a considerable scale, all the more unacceptable as the disadvantage is often to be found in a sea of relative prosperity. Cloke *et al.* (1994), for example, found that in 1990 about 20–25 per cent of England's rural households were living in poverty (admittedly just one element of 'disadvantage') and a host of other measures highlight substantial minorities suffering pronounced 'disadvantage'.
- it implies the *squandering of a considerable resource*, namely a large share of rural Britain's human capital. If, as we argued in Chapter 4, much of rural economic development involves adding value to neglected resources, then an argument that is valid in relation to redundant factories or poorly managed landscapes is at least potentially applicable to excluded people as well. The French in particular were quick to see this, and by 1995 had coined the term 'insertion development' to denote the pursuit of local development by the explicit 'valorisation' of excluded people living in that locality (ADEFPAT, 1996; Brunet, 1996).

## SOME KEY ISSUES

Excluded from what? One rather sweeping answer to that question is 'from the rights, benefits and opportunities taken as the norm in contemporary society', a phrase used in an evaluation of an EU programme to combat exclusion in Ireland (Harvey, 1994). More pointedly, most commentators point to around eight such rights, benefits or opportunities:

- *an adequate income* – certainly sufficient to afford the basics of life, and at least occasional luxuries, without undue stress or hardship;

- *employment* – not just as a means of acquiring that adequate income, but as a source of self-esteem, social networks, personal satisfaction, daily routine and other intangibles;
- *affordable and adequate housing*, which is crucial both to the quality of life of individuals and households and to the vitality of whole communities (see Chapter 6);
- *education and/or training* – to raise horizons, widen the range of opportunities and possibilities, and enrich lives;
- *information and advice* on a host of matters including job and training opportunities, welfare entitlements, and healthcare;
- *easy access to services and facilities* that many take for granted – for example a range of shops, healthcare, post offices, pubs and transport to go further afield (see Chapter 8);
- *social integration* – the ability to play a full part in the life of the local community, to have support networks, to be if so wished, a 'normal member of society'; and
- *the power to influence events* that affect their well-being – to be the 'subject' of their lives and not just the 'object' of other people's.

## WHO ARE 'THE EXCLUDED?'

There has been a good deal of research into who most suffers exclusion from such 'rights, benefits and opportunities' (for example Cloke *et al.*, 1994; Shucksmith *et al.*, 1994, 1996; Beatty and Fothergill, 1999; Shucksmith, 2000 and the various research reports cited therein). Significantly, most are people who are *detached from the labour market* for one reason or another, though people in low-paid work and many self-employed certainly do comprise other significant groups.

Most of these 'detached' people fall into one *or more* of the following groups:

- elderly people living alone and elderly couples;
- disabled people and the long-term sick – many of the latter being men over 55 years of age;
- unemployed people, especially the long-term unemployed;
- lone parents;
- young people under 25 who are members or potential members of only a local, rather than national, labour market; and
- women generally, but especially those who are carers.

Of these, the most numerous are elderly people largely reliant on state benefits.

An important point is that just as the disadvantage is often compounded for those who fall into two or more of those categories – for example unemployed, lone parents or elderly carers of their elderly partners – so many of the elements of disadvantage often afflict the same person or household; in other words they suffer 'multiple deprivation'. For example, many young people in rural areas are caught in a cycle of intermittent, poorly paid jobs, periods of unemployment, a lack of the education and training that would improve their employment prospects, poor accommodation and the lack of a car that would allow access to a wider range of job opportunities. Such is 'exclusion'.

### WHAT MECHANISMS EXCLUDE THEM?

As Shucksmith (2000: 12) put it, to understand the prevalence of 'disadvantage' in rural Britain we must look at the 'ways in which resources are allocated in society – through market processes (for example payment for work), through transfer payments and services provided by the state, through collective actions organised via voluntary bodies and through ... processes associated with networks of families and friends'.

There are various ways of categorising the ways in which resources, status and power are allocated in rural society, but the following provides one approach:

#### *The labour market*

At the heart of much disadvantage and poverty is the simple fact that in rural Britain there are not enough jobs and, more particularly, not enough secure, well-paid jobs, taking the latter to mean jobs sufficiently secure and well paid to provide an incentive to come off benefits and to meet childcare, transport and other resultant costs. This 'labour market' problem derives in part from the restructuring and job-shedding that has afflicted many industries, both nationally and, more specifically, in rural areas, especially agriculture and related businesses, but it stems from other factors as well. Much employment in rural areas, for example, is seasonal and often casual – notably in agriculture, food processing and tourism. Much of it is low-paid. And the 'support infrastructure' to help disadvantaged people, often with little or no capital, no car and a poor level of education, to get into employment is often insufficient – transport to work, affordable childcare in the right location, access to training and access to information on all of these matters being often highly problematic.

#### *The housing market*

The two key points here are that in rural areas demand for owner-occupied housing greatly outstrips supply, with obvious effects upon prices, and that the stock of 'social housing' – meaning those dwellings explicitly reserved for lower-income households and let to them at a discount – is in seriously short supply. In the owner-occupied sector the demand/supply mismatch is easily explained – strong demand fuelled both by 'household fission', meaning the growth of one- or two-person households, and by largely middle-class in-migration – and constrained supply because of restrictive planning policies and the NIMBY ('not in my back yard') attitudes often most apparent in those who have recently bought into the rural idyll. The result is that even *middle*-income households without a significant deposit struggle to buy in much of rural England; the truly 'disadvantaged' have virtually no chance. As for social housing – now comprising only 14 per cent of the rural housing stock compared with 23 per cent in the towns – the stock was severely eroded by the sale of council housing in the 1980s and 1990s and was never properly replenished. This was for a combination of reasons including much the same 'anti-development' sentiment referred to above as well as the inadequacy of government funding of the housing associations now charged with building the lion's share of social housing in England.

In the language of exclusion, then, planning controls have become instruments of exclusivity, as has the inadequacy of funding for social housing. At a deeper level, the excluders are the NIMBYs and the more comprehensively exclusionary BANANAs ('build absolutely nothing anywhere near anyone') who can, together, set the political climate in which planning authorities operate.

#### *Imperfections of the welfare state*

Here the main point of concern is the often poor take-up of benefits in rural areas by those who are perfectly entitled to them. The problem seems to be a combination of ignorance and confusion about entitlements – an indictment of the complexity of provision and of the availability of good-quality information and advice – and a culture of self-reliance and spirited independence which leads many rural people to struggle on in hardship rather than ask for help.

#### *Deficiencies in the delivery of services*

Rural areas provide notoriously difficult 'terrain' over which to deliver services to disadvantaged people, whether they be public sector (such as much of healthcare), private sector (such as the retailing of food), voluntary sector (some social care and advice services) or multi-sector (to include transport and training). This is because the groups identified earlier as typically disadvantaged tend not to have cars to reach distant 'outlets' easily, and because the delivery of services to geographically dispersed settlements and households is generally expensive and therefore curtailed (The next chapter is wholly devoted to this core issue.)

#### *Inadequate social and community support networks*

Finally, several factors have tended to erode the supportive, close-knit local communities which – without ever wishing to glamorise rural life before the arrival of the 'affluent society' or to forget that such communities could invade privacy as well as offer support – often did, and do, provide a source of assistance to disadvantaged rural residents. Included among such factors are the growing number of women who now work full-time, often some miles away, behavioural and cultural changes linked to the very widespread ownership of cars and television, the rapid turnover of people living in a particular place as migrants come and go and the fragmentation of the extended family. The last point is important; hard data are limited, but it seems certain that the number of *three*-generation families – those comprising children, parents and grandparents – now living in close proximity is vastly less than in the past and that all three generations are often the poorer for this in times of adversity. Again, the serious shortage of affordable housing is partly to blame, but there is, of course, a host of other contributory forces, both social and economic.

### TWO FURTHER QUESTIONS

Before turning to the 'toolkit' for trying to alleviate some of these problems it is essential to ask two further questions: 'Do they have a truly *rural* dimension?' and 'Do they have a *local* dimension?' The answer to each question is 'yes', but that neither dimension should be exaggerated. In other words, most disadvantage and exclusion is essentially aspatial, with its causes linked to the national and/or capitalist context and to major contextual forces such as globalisation.

But there *is* a truly rural element – or, more accurately, there are four. First, the plethora of small, scattered settlements which generate difficulties of accessibility and service delivery not encountered to any significant extent in the towns. Second, a disproportionate endowment of certain land-extensive activities, notably agriculture and tourism, that generate particular labour market problems already referred to. Third, rural areas are by and large deemed more desirable than the cities and suburbs as places to live by many of those who can afford to make the choice, and some of the social and economic implications of their making this choice have also been alluded to. Fourth, despite strong forces tending towards the homogenisation of our national culture – meaning in this context the diminution of urban/rural differences in attitudes, values and ways of life – some broad differences do seem to persist. Most relevant in the present context are a greater tendency in rural areas to harbour a spirit of self-reliance among those who might otherwise seek out support from the state, and, linked perhaps to the nurturing of the 'rural idyll' in the hearts of those who have come to live in rural areas, a limited acceptance that in this 'green and pleasant land' real social problems actually exist among their neighbours. All four of these 'rural dimensions' to social exclusion have to be recognised in the fashioning of policy responses.

And is there a *local* dimension to exclusion? Again, 'yes', for the simple reason that areas within rural Britain vary considerably in both their needs and resources. In the generally prosperous Cotswolds area of Gloucestershire, for example, finding affordable housing is for most disadvantaged people a much greater problem than getting a job. But in the deprived coalfield areas of Nottinghamshire or County Durham, where house prices are much lower and unemployment significantly higher, the reverse is the case. And, on the 'resource side', one area might be well-endowed with community-based organisations with a readiness to get involved in regeneration initiatives, while another might seriously lack this resource. Such disparities provide one – not the only – justification for a local dimension to policies and programmes for social inclusion.

## THE TOOLKIT

As already explained, much if not most of the action needed to address social exclusion in rural Britain has to be taken at national level; it relates to macro-economic and social welfare policy, and measures such as the national minimum wage, the New Deal and other attempts to facilitate a movement from 'welfare to work' are of that kind.

But a great deal can be done at the local level – by the local authorities, health authorities, local development agencies, the local voluntary and community sector, for example, often working in partnership. Here we attempt a brief overview of some of the steps than can be taken and that figure in many local strategies to promote social inclusion in rural Britain. Because full-time, secure employment is often an effective route out of poverty, many involve replacing 'detachment' with 'attachment' to the labour market and these are considered first.

### EMPLOYMENT INITIATIVES

Stimulating the local economy is perhaps the single most effective action that can be taken by a local development agency seeking to promote social inclusion. This can relate to the creation or attraction of new firms, the support of existing businesses, the creation of jobs in the public, voluntary and community sectors, including 'community enterprise', to the encouragement of local purchasing etc. (See Chapters 4, 5 and 9 for more on the stimulation of the local economy.) The point is that getting more jobs and better jobs in an area is almost always a valuable tool of local *social*, as well as economic, policy.

### AIDING THE TRANSITION TO WORK

But the above measures, while necessary, are rarely sufficient to ensure that a good proportion of the newly created jobs are taken up by local excluded people. Complementary measures are generally needed including:

- the adequate provision of information about the job market and how it might be entered or re-entered;
- the provision of childcare of a quality, in a location and at a price that persuades parents at home with children that it is viable and beneficial to take a job;

- the provision of training and if necessary remedial education, again in a location and in a manner that is genuinely accessible; and
- the provision of transport so that the place of work or of training, if appropriate, can be accessed on the right days and at the right times.

That is not to say, of course, that none of this is the responsibility of the excluded person himself or herself, but experience shows that particularly in rural areas, for reasons given above, such hurdles can seem insuperable without some external assistance. There is no space here to spell out the range of actual ways in which the information, childcare, training and transport might be actually provided in a rural context – for some suggestions see Case Study 14 and other chapters, notably those dealing with involvement, accessibility and enterprise – but the guiding principle must be imagination and flexibility; such is the requirement of rurality when it comes to service delivery.

### *AFFORDABLE HOUSING*

The challenge here concerns how to increase the stock of dwellings effectively available to low-income people living in the area. A great deal has been written on this subject in recent years (e.g. Rural Development Commission, 1999) but in many respects it comes down to a need to assemble five crucial ingredients in the same place at the same time:

- clear, quantified evidence of the amount and nature of local housing need;
- land at a price considerably lower than that pertaining to 'development land' because of the operation of the land-use planning system;
- political support at both the local authority level where prime responsibility for land-use planning and social housing lies, and at the very local level where NIMBY opposition can be rife;
- development machinery, which generally means a housing association or other 'social landlord' ready and willing to build, and then to manage, a few houses on often small and remote sites; and
- finance, which generally means a sufficient commitment to a housing association by the Housing Corporation and/or local authority.

Getting those five elements in place simultaneously in a particular location can prove more elusive than getting recalcitrant sheep into a small pen in the middle of a large field. And a whole new profession has arisen – that of the 'housing enabler', whose job it is to play the role of benign sheepdog. Space prevents a consideration of all that can be done at the local level to facilitate this process – from sponsoring local housing needs surveys to setting up 'land banks' so that supportive landowners can pledge cheap land while the 'sheepdog' gets to work, and ensuring that planning strategies incorporate provision for approving social housing on so-called 'exceptions land' which will be relatively cheap since any other kind of development is ruled out there. But in much of rural England no local, social inclusion programme can be taken seriously if it fails to address the supply of affordable housing.

Of course, it is not self-evident that increasing the supply of affordable housing necessarily means new-build, though that was the almost universal interpretation of 'the solution' through the 1980s and 1990s; much might be achieved, with considerably less angst, by focusing on the potential of the *existing* stock of dwellings regardless of its ownership, and rendering some of it more accessible to the excluded.

### COMMUNITY DEVELOPMENT

In various ways, building up the capacity and vitality of local communities can significantly improve the lot of the socially excluded. In relation to poverty, credit unions and LETS (Local Enterprise Trading Systems) schemes can help without a need for the excluded to find a new income source. In relation to paid employment, informal support networks can help with childcare, car-sharing or lift-giving schemes and word-of-mouth information about opportunities. In relation to housing, a vital and active community can erode NIMBY opposition and even secure promises of low-cost land by generally raising awareness and understanding of the problems, perhaps through undertaking a village appraisal and encouraging a wide discussion of its findings. In relation to services, local campaigns to retain and improve facilities, ranging from post offices to village halls and public transport, can all help excluded people who are disproportionately dependent upon them. And finally, attempts to bridge the gap between long-term residents and incomers – many of whom will have both assets and problems of their own, such as lonely mothers with workaholic husbands, young children and no local friends or relatives – can be to their mutual advantage. On all of this see other chapters, especially those on sustainability, community, accessibility and involvement.

### INFORMATION AND ADVICE

We have seen that a good deal of exclusion arises from, or is sustained unnecessarily by, ignorance or a lack of appropriate advice – on job or training opportunities, benefits entitlements, debt management, childcare, transport availability and so on. The lesson seems to be that *ad hoc* initiatives in this regard are inadequate; an integrated, area-wide strategy is needed. Such strategies need to embrace different 'sectors' or agencies, including the local authorities, the Benefits Agency and the Citizen's Advice Bureau, as well as different 'delivery mechanisms', including well-located advice bureaux of various kinds and also mobile units, peripatetic or outreach workers, local outlets such as post offices and schools, and telecommunications from the humble telephone upwards.

What is important is not just the dissemination of information and advice to the *sufferers* of exclusion; we must not forget the potential alleviators of exclusion such as parish councils, which could do much to facilitate inclusion by promoting many of the initiatives listed in the earlier paragraphs, and school governors who might be persuaded to host after-school childcare or 'breakfast clubs' at their school. Many, of course, already do so; the task is to spread good practice, and that means spreading information and advice, as well as using subtle persuasion.

### *AREA DEVELOPMENT PROGRAMMES WITH SOCIAL INCLUSION 'CENTRE-STAGE'*

So much for the local 'inclusion toolkit'. The first case study below looks at the main 'inclusion tools' deployed in one rural development programme – Dorset. It might also be useful to speculate on the content and character of an area-wide rural development strategy that genuinely put social inclusion as its overarching objective (LEADER Observatory, 2000f). Such a strategy would

- treat excluded people as a resource as much as a problem, and as the subject, not just the object, of local development. The resource would be researched and quantified and the constraints on adding value to it (see Chapter 4) carefully explored;
- build on an elaboration of 'what are the needs of excluded people?' and 'what are their capabilities?', to try to put the two together in specific projects. One example might involve more active elderly people participating in social car schemes or in after-school care; another would have unemployed young people providing computer literacy classes for their elders;
- involve excluded people and their representatives in the planning and delivery of the development programme, perhaps through adequately resourced consultative groups including young mothers, activists in support groups for carers, lone parents and claimants;
- promote sustainability at the very local level, with an emphasis on internalising the supply and demand for local goods and services within the village, parish or small town;
- promote human resource development, capacity building and the caring community at the local level;
- aim to increase the amount and range of community-owned assets that give local communities an income and a measure of power – community businesses, buildings, resource centres, development trusts, exchange systems; and
- ensure the 'inclusion proofing' of all the various projects in the programme – for example by making grants to a village hall committee conditional upon a prior survey of the whole community's needs and not just the wishes of an inner circle.

In these and other ways it would be possible to bring social inclusion centre-stage in a local development programme.

## CASE STUDY 13: THE DORSET RURAL DEVELOPMENT PROGRAMME 1994–98

The Dorset RDP was launched in 1994 by a local partnership of public and voluntary bodies to promote economic and social development across a large swathe of rural Dorset with 140,000 inhabitants. By mid-1998, when the present author undertook an evaluation of its progress (Moseley, 1998), some 70 projects had been supported at a total cost of about £9 million. 'Social inclusion' was just one of a number of objectives pursued, others being more economic or environmental in character, but very few of the 70 were totally irrelevant to that concern. It proved impossible to quantify the amount of 'inclusion' achieved by the programme but column 3 of Table 7.1, which lists the 17 projects most closely aimed at social inclusion, suggests that the contribution was real enough.

The collective cost of those 17 projects over the three years approached £1 million – the 'big money' going on the provision of serviced industrial land and workspace in five small towns and on the restoration of Swanage Pier to attract tourists – all projects improving local job prospects and, hopefully, 'inclusion'. Three conclusions might be usefully drawn: first, that addressing social exclusion involves a great variety of projects; second, that there is a danger of spending limited resources on 'gap filling' rather than on stimulating an upward spiral of development (a difficult distinction to draw, but compare the 'developmental' mobile training centre (no. 5) with the 'gap filling' funding of an extension to a day centre (no. 13); and third, there is a danger of what might be termed 'project-ism' – meaning construing the promotion of 'development' as simply the funding of well-chosen projects rather than that *plus* the influencing of key outside agencies (a point discussed more fully in Chapter 13).

**TABLE 7.1** ***The Dorset Rural Development Programme: projects with a strong 'social inclusion' element***

| Project | What it involved | Relevance to social inclusion |
|---|---|---|
| 1. Community workers serving Portland – 13,000 population – and the small town of Sturminster Newton plus its hinterland | Two workers who helped launch a Volunteer Bureau, summer play schemes, after-school clubs, support groups for elderly and unemployed people etc. | Childcare meant mothers were more able to take paid employment; also the social isolation of excluded groups was reduced and support networks enhanced |

*(Continued)*

**TABLE 7.1** ***Continued***

| Project | What it involved | Relevance to social inclusion |
|---|---|---|
| 2. Swanage Advice Shop for young people | A full-time development worker, advising/signposting young people (aged 14 to 25) on matters of training, employment, transport, access to services, health and personal concerns etc. | Some success in helping young people get jobs and training. The advice shop also spawned a 'Train and Build' programme which gave six people training, employment and then their own self-built accommodation |
| 3. Maiden Newton Youth and Community Centre | Repair of, and better facilities in, a village centre, especially to provide a better meeting place for young people and for the wider community to meet and socialise | Some new activities e.g. a playgroup and an after-school group helped mothers to work part-time. More generally, the project helped to develop the youngsters' social skills, and also community development and the strengthening of informal support networks |
| 4. Sherborne Citizen's Advice Bureau – home visiting service | By home visiting, a professional advice/advocacy/representation/negotiation service for people with transport difficulties within a ten-mile radius | Client monitoring indicated success in reaching disadvantaged people, for example low-income households with debt problems. Overall, helped boost benefit take-up in that area. |
| 5. Mobile training centre | A caravan converted into a training centre, plus full-time manager, spent a few days at a succession of rural venues. 'Taster' courses, courses for women returners, variety of IT courses etc. | In use 200 days a year with an average of ten trainees per day and 500 different individuals per year. Successful in reaching rural residents for whom distant urban training facilities were largely inaccessible. Especially useful for potential women returners |

*(Continued)*

**TABLE 7.1** ***Continued***

| Project | What it involved | Relevance to social inclusion |
|---|---|---|
| 6. Rural transport adviser | Promoted and developed community-based transport schemes by disseminating information on good practice and supporting individuals, communities, voluntary groups, transport operators etc. | Indirectly alleviated social isolation of many rural residents without a car. Some help with journey to work and journey to training trips but other trips were quantitatively more significant |
| 7. Wincombe Employment Centre, Shaftesbury | Twelve small business units (34 jobs). Also 54 childcare places and an outpost of an FE college providing training | Childcare helped working mothers. Training improved employment prospects especially of young people |
| 8. Community projects fund | A small grants scheme, across the whole Rural Development Area supported 50–60 projects per year (average grant £500). All projects were locally originated, run and co-funded and for local people | Projects often related to community facilities and equipment, children's play, recreation. No explicit targeting on disadvantaged people but outputs included empowerment of 'ordinary village people' and better social/support networks at local level. |
| 9. Cold Harbour Business Park, Sherborne | A conventional small 'industrial estate', but one which opened up a 'landlocked' site for 54 units of social housing | Enabled a valuable addition to the local stock of social housing |
| 10. Work 'n' Play initiative | Employed a worker to develop, across the area, out-of-school childcare schemes, training provision in play-work, and a play club 'network' | Built up the childcare infra-structure to allow more mothers to enter the labour market |
| 11. Rural families project | Part-time worker, serving three parishes, to enhance family support services and foster quality day care and leisure opportunities for older children | Support for families in difficulty, improved community networks, better childcare |

*(Continued)*

**TABLE 7.1 *Continued***

| Project | What it involved | Relevance to social inclusion |
|---|---|---|
| 12. North Dorset disability information service | Support for the North Dorset disability information service | Disabled people and carers had access to specific information on disability |
| 13. Bridport day care centre | An extension built on to the day centre and improved facilities for its clients | Better quality of life for elderly people, including a luncheon club, chiropody service etc. |
| 14. YMCA upgrade – Portland | Multi-purpose centre for young people including hostel, social, training, advice and counselling provision | Better social integration of young people in a deprived area of Dorset |
| 15. Bridport Foyer | Conversion of a former youth hostel to provide a secure home with training and counselling for young people and to help them get work experience | Sheltered housing, support and improved job prospects for disadvantaged young people |
| 16. New Opportunities for Women | 14-week training course in confidence and various vocational skills for women returners – held in a large village and attended by about 20 women | Of the 20 women attending, four progressed to employment and five to further education/ training |
| 17. Computer training for disabled people in Maiden Newton and Sturminster Newton | IT and other courses to enhance employability of disabled people. A 'job club' was set up to aid job search | There were seven initial trainees in Maiden Newton of whom three found employment within a few months |

## CASE STUDY 14: THE PEAK DISTRICT RURAL DEPRIVATION FORUM

The Peak District Rural Deprivation Forum is a voluntary organisation established in 1992 to raise awareness of the nature and extent of deprivation in the Peak District and to work with other agencies and local people to develop solutions.

Its geographical focus is almost exactly that of the Peak District National Park, an area of some 1300 sq. km with about 40,000 residents. The geographical context is important as the area is almost literally the 'back garden' of several million people in nearby towns and cities, a proximity that has encouraged a mass of largely middle-class commuters and retired people to come to live in its idyllic environment and 22 million others to visit each year. Almost all of these people would generally strongly oppose the area's 'development'. Also significant is the fragmentation of the area politically and administratively, spanning as it does parts of three counties, four unitary authorities and five districts. Only one body, the National Park Authority, focuses exclusively on the area as a whole and its remit and ethos relate essentially to conservation and recreation.

About ten years ago, an influential research report (Scott *et al.*, 1991) revealed this beautiful rural haven and playground to also have serious social problems. It highlighted the poverty experienced by many elderly people, lone parents and those engaged in part-time and/or low-paid employment; the very limited range of job opportunities; the social consequences of house prices inflated by middle-class in-migration and restrictive planning policies; and the isolation, loneliness and frustration of many people who often felt strangers in their own villages. And it stressed that this deprivation was largely hidden from sight by the beauty of the area – so different from the neighbouring coalfields and industrial areas – by the preconceptions of those who came to live, work or recreate there and by the fact that those in greatest need were often the most reluctant to admit it.

The publication of that report – one of the first for any British rural area to be based largely on the qualitative analysis of local people's testimony – led to a series of 'open forum meetings' at which various statutory and voluntary agencies and concerned individuals considered how best to respond. Those meetings led to the creation of the Forum which has been neatly characterised by Scott as 'neither top-down nor bottom-up, but middle-across' (Peak District Rural Deprivation Forum, undated: 7). By that he means that it is a loosely structured network composed largely of middle-ranking activists in those agencies, albeit one that has expanded and broadened in recent years to enjoy a membership of some 400 people and organisations, many of them of a more 'grassroots' nature.

The work of the Forum has effectively fallen into two phases – roughly 1992–97, and thereafter. The first phase, greatly helped by three years' funding from the Rural Development Commission and the North Derbyshire Health Authority, comprised a programme of further research, awareness raising, networking and campaigning. More specifically, a number of working groups were established, typically with a dozen or so members drawn from such bodies as the area's Council for Voluntary Service and its Rural Community Councils, the local authorities and health authorities, welfare rights groups, carers associations; and the worlds of community education, mental health, counselling, advocacy, transport and housing. These groups focused on transport and accessibility, low income and welfare rights, health and community care and other broad themes, gathering further local evidence and producing and disseminating reports, and in one case a video and training pack, all designed to help a range of agencies and community groups better appreciate the 'rural deprivation' dimension to their work.

Those endeavours, and associated conferences, newsletters and campaigns, did have some effect in raising awareness and changing attitudes. An evaluation undertaken in 1998 (Scott and Russell, 1999) noted that over the previous six years new coalitions had been formed such as the Peak District Transport Partnership, the Peak Rural Welfare Rights Project, and the Voluntary Sector Infrastructure Working Group – all with the intention of better targeting resources on the more deprived. The evaluation also detected modest but useful shifts of resources or changes of emphasis in at least one community care plan, the inclusion of anti-poverty statements in various strategic documents and the 'gingering up' of certain voluntary bodies hitherto somewhat skating round the edges of what the Forum deemed to be the key issues.

There was also the point that individual members of the Forum and its working groups had been encouraged and emboldened by their experience. 'A number of projects ... such as a Patients Transport Scheme and establishing Citizens' Advice Bureaux in GP surgeries, have been the product of individual members drawing strength and ideas from the Forum and then developing projects themselves' (Scott and Russell, 1999: 29).

Then, around 1998, the emphasis changed, partly from a wish to move forward from awareness-raising to project delivery on the ground and partly from a sense that 'ordinary local people' had hitherto been directly involved in the Forum's work to only a limited extent. 'Is it possible to integrate the efficiency of the professional with the experience of the local?' asked the Forum's director, Denise Servante, at one of the seminars convened to reassess the Forum's direction, and some projects launched since 1998 are an attempt to respond positively to that question.

One such initiative has been the Focus Group, described as 'an informal friendly group that anyone in the Peak District is welcome to join' which meets monthly to 'find out about the main problems of living in the area from those with first-hand knowledge'. The deliberations of this group have already led to successful representations to local Citizens' Advice Bureaux and transport operators. Another initiative has been the Small Groups Project, funded by Comic Relief, which offers support – for example in relation to funding, IT provision and financial management – to a host of very small voluntary groups in the area which involve marginalised people.

Most important, however, has been the Amethyst Project, supported over three years by a consortium of eight funders both national and local, to help groups of local women identify and overcome problems that prevent them from reaching their full potential. The idea is that project officers respond to agendas articulated by disadvantaged women themselves, and five groups, comprising about 50 women in all, now meet regularly to discuss and take forward practical proposals relating, for example, to parenting, road safety, local transport and learning computer skills. All the indications are that the project's 'soft outputs' – relating to the personal development of the disadvantaged and often 'forgotten' women involved – are at least as significant as other tangible successes relating specifically to such issues.

And so, almost ten years from its inception, it is possible to conclude that the Forum has demonstrated the value of a single-issue ('deprivation'), area-focused (Peak Park) network organisation linking the statutory, voluntary and local community sectors in an endeavour to better understand, communicate and resolve hitherto hidden or ignored social problems. The key challenge has been to work *with*, and not just *for*, excluded people. This has taken time, care and perseverance but some significant successes have been achieved.

---

***SEE ALSO …***

Case Studies 15, 16, 17 and 21 each relate to the alleviation of social exclusion.

***FURTHER READING***

See Shucksmith (2000) for an excellent, research-based overview. See also Policy Studies Institute (1998) and Countryside Agency (2000a) and the local studies cited in each of these.

# 8

# Accessibility: Bringing Within Reach

A horse! a horse! my kingdom for a horse!

Shakespeare, Richard III

## DEFINITIONS AND IMPORTANCE

The accessibility problems experienced by many people living in rural areas have been a key concern of rural development programmes. *Accessibility* may be defined as 'a measure of the ability of people to reach the things important to them', but that definition is rather too broad for our present purpose. In this chapter we are concerned with 'people's ability to reach things' only in its spatial or geographical dimension. Of course, geographical space is only one constraint upon people's accessibility in the broader meaning of that term, as Chapter 7 on the sources of social exclusion made clear. But the 'rationing mechanisms' implied by other factors, such as the uneven interpersonal distribution of income, education, training, information and social skills, are outside the scope of the present chapter. Our concern here is with how space constrains people and with what might be done about it.

Accessibility is not to be confused with *mobility* – the ability of people to move around. It is possible for someone to enjoy a good deal of personal mobility but still have many of the 'things important to them' effectively outside their reach – a two-car household on an even quite large Hebridean island provides an example. Conversely, a house-bound person can enjoy considerable accessibility if everything he or she needs is delivered to the door. In other words the location of the desired 'thing' – and indeed the residential location of the 'person' who has the desire – plays as significant a part in his or her accessibility as does his or her mobility. That statement brings us to the simple but still valid conceptualisation of 'accessibility' proposed by the author some years ago (Moseley, 1979); accessibility in the spatial dimension has three elements – *people* needing a *link* to an *activity* – each of which may be properly subject to public policy. In short, there is a lot more to accessibility than transport.

Using such very general words as 'things' and 'activities' to denote what the rural resident wishes to access was necessary above because those 'objects of desire' are so varied. Included are services such as healthcare and retailing, facilities such as hospitals and sports grounds, employment, training provision, social contact with friends – in short, anything important to the person in question which is normally located some distance away.

The challenge, in policy terms – the challenge that any agency concerned with the social aspects of rural development must inevitably face – is as follows:

- how to expand the effective personal space of people suffering 'accessibility deprivation' such that they can more easily reach the desired 'things'; and/or
- how to bring within their personal space more of the things that they desire.

The first of these options is a matter of transport policy; the second is a matter of service delivery and the geographical distribution of whatever is desired.

No-one would argue that physical accessibility is never a problem within towns or on large suburban housing estates, but it is *inherently* and *seriously* a problem in rural areas which contain, by definition, geographically dispersed people, services and settlements.

All this has come to the fore over the past 40 years for two main reasons. First has been rising car ownership such that today about 46 per cent of Britain's rural households have one car and 37 per cent have two or more (Countryside Agency, 2000a). This has eroded both rural public transport and local small-scale outlets such as village shops and schools. Second is the greater ability today of many service delivery agencies, from the major food retailers to the health authorities responsible for primary and secondary healthcare, to reap 'economies of scale' and thereby to achieve more sophisticated provision by concentrating service delivery in a few large and necessarily widely spaced outlets – namely, in this example, superstores and district general hospitals respectively. The 17 per cent of rural households without a car, and some members of car-owning households, have lost out as a result.

There is a further key development which has yet fully to show its hand, namely the continuing revolution in information and communication technologies (LEADER Observatory, 1999b). In the challenge to enhance the personal accessibility of rural residents, that revolution seems likely to prove a double-edged sword. In some respects it is already eradicating distance – for example by delivering banking services into people's homes. But it will serve also to increase social isolation by further eroding the number of information-based village and small-town outlets, including banks, post offices, police stations and local government offices. And, once again, it is the two- and three-car households who can best shrug off such effects, and the no-car households who will suffer.

## SOME KEY ISSUES

This is not simply a 'social welfare' problem. It is a broader rural development issue because it affects the ability of local areas to sustain the development processes described in other chapters. Three examples will suffice to make this point.

First, we have argued in Chapter 2, and will reassert in Chapter 10, that an area's social capital is a crucial ingredient of its development. But that capital will be seriously reduced if there are fewer and fewer 'venues' such as pubs, churches, shops and schools where people can interact and gel as a 'community', rather than exist as a collection of individuals living close to one another. Second, employers need labour, of course, often skilled labour, and they cannot afford to ignore those people – especially women, young people and the unemployed – who are least likely to have their own transport to get to work and to training opportunities. Third, to lose local services and the implied 'local spend' is to sacrifice the local multiplier; recuperating from an operation in a big-town 'district general hospital' rather than in a small-town 'community hospital', for example, does nothing to sustain the rural economy.

Each of the other key issues considered below relates to the complexity of responding to these problems – and 'complexity' is certainly the right word. It is clearly tempting to address aspects of the problem in isolation, for example by launching a programme to help small-shopkeepers to invest in the new equipment that might help them survive in a village location, or a programme to make mopeds available to young people needing to get to jobs or further education colleges in distant locations. But the challenge is to take a holistic view; certainly to consider simultaneously not just transport but also the location and availability of the services and facilities in question.

Going on from that is a need to consider in the round all the *different modes of transport* potentially available – scheduled bus and train services; bus services provided under contract such as schools transport; demand-actuated public transport such as 'ring and ride'; taxis and non-emergency ambulances; minibus services often reserved at present for just one client group such as people with a particular disability; and the private car. The last-mentioned should certainly not be overlooked as a resource for 'good neighbourliness', given the importance of 'social car schemes' operated by volunteers and organised car-sharing, especially for journeys to work. In addition, there are mobile and delivery services such as mobile shops and libraries and the ubiquitous post van.

Moving on to the provision of *fixed location service outlets*, such as shops, pubs, churches and post offices, the challenge in each case is to resolve what might be termed *the triangular dilemma of rural service delivery* (Moseley, 2000c). The dilemma concerns how society might strike the best possible compromise between three worthy but competing goals: high service quality; low unit costs arising from economies of scale; and a good geographical spread of

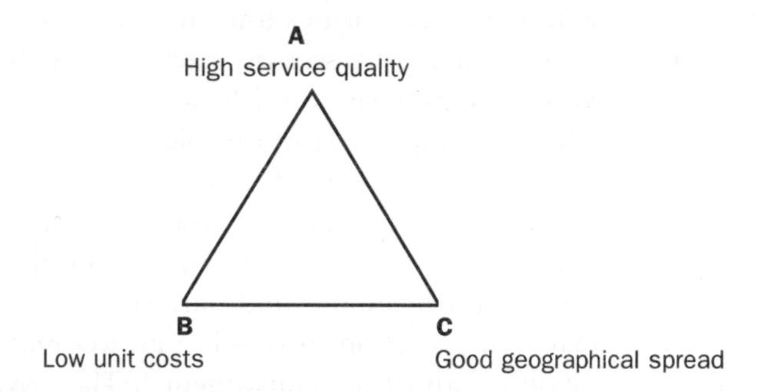


outlets, such that the less mobile rural residents are not unduly excluded. With reference to the diagram above, this dilemma may be clarified by giving examples of service outlets that generally achieve just two of those three *desiderata:*

| | |
|---|---|
| A & B but not C | primary health care centres |
| B & C but not A | most branch GP surgeries |
| A & C but not B | most small primary schools |

It is tempting to suggest that much of the tension and resultant innovation in rural service delivery reflects the frustrations and challenges of trying to achieve a happy mix of A, B and C through a process of lateral thinking, given that A, B and C are fundamentally incompatible.

This call to look at transport provision in the round and to grapple systematically with the triangular dilemma of rural service delivery, underscores a need 'to manage the accessibility system' at an area level. In England it is the county councils and unitary authorities that have the statutory responsibility to co-ordinate the provision of conventional bus services as well as responsibility for the transport both of schoolchildren and social services clients, and, of course, they also have the strategic land-use planning responsibility. Therefore it is they who seem best placed to take the lead and to link up with healthcare transport provision, the various mobile service operators and other actors.

But experience, some of it reviewed in the two case studies below, suggests that whole counties tend to be too large to achieve this co-ordination effectively and recent years have seen many attempts at the *sub*-county level to forge partnerships, involving the public, private and voluntary sectors, to manage the delivery of transport. Thus 'accessibility', like so many other issues reviewed in this book, has come to be addressed principally at an intermediate level somewhere between the individual settlement or parish and the wider county or region.

Of course, the size of the appropriate geographical unit is not the only issue in achieving better co-ordination; much thought in recent years has also gone into the proper division of responsibilities between the state (using that term loosely to include the local authorities and the 'quangos'), the private sector and the voluntary and community sectors. Each, clearly, has a role to play in the jigsaw of accessibility provision and local area-based partnerships often provide a good vehicle for exercising those roles.

With the emergence in recent years of new patterns of local governance, the general consensus on 'who does what' in accessibility provision has evolved along the following lines. The state, including the 'local state', provides less directly than hitherto but it facilitates and purchases more, striking a balance between the standards of provision to which it aspires and those which it can afford to purchase on behalf of its 'constituents'. The private sector (the rail and bus operators, taxi firms etc.) then sells its services to two clients – the general public, as fare-paying passengers, and 'the state' as guardian of the public interest as described in the previous sentence. The voluntary sector, whether largely informal in style – as with the charities that typically organise and run social car schemes whereby volunteers take frail and disabled people to hospital and elsewhere – or more formal – as with the trusts that now run most ambulance services – operates partly under contract to the state and/or in pursuit of its own charitable purposes, drawing its funding from any available source.

In all of this, an important question concerns how to trade off the goal of 'national standards' – the ideal that wherever one lives the pattern of opportunities should be more or less the same – with that of respecting local diversity. For example, with regard to the mail collection and delivery service we have in Britain, strict national standards apply regarding both the cost borne by the client and the quality and frequency of provision; a daily delivery of mail at a standard tariff is a right enjoyed by virtually everyone regardless of where in the UK they live. But with passenger transport there is no such right; in that case local diversity is the norm, with the local authorities generally at liberty to strike the balance between cost and service quality as they see fit. The 'local initiative' thrust of this book obviously points to the latter as being generally preferable, not least because it is more likely to produce innovation in the search for solutions and a greater release of local talent both in contributing ideas and in helping to deliver the locally determined 'package'.

## THE TOOLKIT

This brings us to the toolkit for delivering personal accessibility. Recalling the two sides to the challenge spelt out earlier, the task is to enlarge the 'personal space' surrounding people by improving the transport available to them and/or to bring more of what they desire into that personal space. Another option, helping isolated people to move home so that their 'personal space' in the future encompasses more of what they desire – essentially the encouragement of migration out of the most rural and poorly served areas into the better served villages and towns – is not considered here for lack of space. But it was once a key rationale of 'key settlement policies' which sought to steer both residential development and services/jobs to just a few, carefully selected places and which may be re-emerging as town and country planners today frequently do likewise on the grounds of reducing unnecessary car use.

### *THE PASSENGER TRANSPORT OPTION*

In most rural counties this amounts to some form of the 'string vest' policy which involves:

- ensuring a basic inter-town network of good quality transport provision by supplementing, where necessary, commercially viable train and scheduled bus services (the 'string'); and
- complementing that provision with a multiplicity of imaginative and, as far as possible, cost-effective services to plug the 'holes'. These latter services frequently include explicitly subsidised scheduled bus services; ring-and-ride services which reap some of the economies of scale of buses together with the flexibility and customer-responsiveness of taxis; community buses and social car schemes with their heavy reliance on volunteer input; minibuses provided by charities and the caring agencies; postbuses and those school buses which permit passenger transport as an add-on to their principal mission; and a variety of car share and car pool services. The list is a long one and the guiding principles are responsiveness to need and lateral thinking.

### *SUPPORTING WIDELY DISTRIBUTED 'MINI-OUTLETS'*

By 'mini-outlets' we mean, for example, village shops and post offices, pubs, village halls, small schools and small doctors' surgeries, as well as threatened small-town services such as community hospitals and police stations. As suggested earlier, the problem is to be as imaginative as possible in trying to resolve the 'triangular dilemma of rural service provision' – seeking a happy compromise between acceptable cost, high service quality and wide geographical spread. (Reviews of good practice in this respect are to be found in OECD, 1991; LEADER Observatory, 1999e; and Moseley, 2000c.)

Two particular mechanisms are emerging in this regard. The first involves getting two or more services under a single roof, thereby reducing the overhead costs of each. This is the *multi-purpose outlet* approach – post offices in pubs, surgeries in village halls, food shops in filling-stations – and Case Study 6 reviewed the spread, and the pros and cons, of this innovation. A variant of it, of course, involves 'broadening the product or client range' of individual outlets. Good examples of this include pubs serving meals to local clubs of elderly people, schools running IT-literacy classes for adults, and post offices delivering some of the services normally provided by branches of the clearing banks.

The second mechanism involves making more use of *volunteers and voluntary organisations* rather than paid staff and management. This approach involves the local community increasing the viability of small service outlets by providing:

- labour – as with school helpers of different kinds and community shop volunteers;
- management – as with school governors and village hall trustees, each of these roles being much more onerous than hitherto;
- finance – as with fundraising for school activities and facilities or for village hall improvements; and
- premises – as with village halls or church halls.

The challenge here is to encourage the continuing supply of these voluntary resources in service provision – for example by means of small pump-priming grants and the deployment of community development support workers – while recognising that there is a practical, and possibly moral, limit to what 'the community' can do.

### MOBILE SERVICES

Another option is to encourage the delivery of services on wheels. Mobile libraries and shops have, of course, a long history in rural Britain. Less common examples are:

- 'information buses' of various kinds, whether operated by a local authority, a health authority or a voluntary body such as the Citizens' Advice Bureau;
- 'mobile day centres' for elderly people, taking a range of facilities and a caravan for private meetings to a number of village halls on successive days;
- conventional bus services delivering parcels or prescribed medicine for example, to isolated people; and
- 'play buses' extending the play group facility to a rota of villages currently without one.

Again, joint provision is possible, as with the mobile library that also carries a CAB advice worker or, more prosaically, the milkman who cashes cheques for housebound elderly people. (See Moseley and Packman (1983) and Moseley and Parker (1998) for a discussion of the range of possibilities and their pros and cons.)

### TELECOMMUNICATIONS

As already suggested, the quickening pace of the ICT revolution is likely to produce a mixed bag of benefits and costs as far as rural accessibility is concerned. But it is certainly part of the 'toolkit' for accessibility improvement and it is already proving of great value (National Rural Enterprise Centre, 1998; LEADER Observatory, 1999b). For example, ICT is opening up all sorts of possibilities – the area-wide brokerage of transport supply and demand (see Case Study 16); the on-line provision of a wide range of information in small village outlets such as shops and schools; the computerisation of post offices and thereby increased scope for the diversification of their business; and enhanced linkage between neighbouring small schools, allowing them to share some management and teaching resources and so become virtually a single school with classrooms maybe two or three miles apart rather than the more normal 20 yards. The real challenge is to develop ICT so that it is genuinely a community as well as an individual resource – as in those four examples.

Finally, there is the *management dimension* to all of this. As argued earlier, there is much to be said for trying to develop and implement 'accessibility strategies' across areas larger than the parish but smaller than the county. The balance of advantages probably lies with developing strategies and assembling resources at area level and then devolving a measure of power and resource to school governors, GP practice managers, active parish councils and others at the local level, thereby capitalising on the dynamic of local enterprise and diversity. This suggests a management and decision-making structure that emphasises local partnership, contracting, facilitating and enterprise, which is, of course, easier said than done.

## CASE STUDY 15: ACCESSIBILITY AND CARE IN THE TEWKESBURY AREA

This case study reports some research undertaken in 1995 for the Tewkesbury Locality Planning Team, a consortium of Gloucestershire Health, the county social services department and other agencies providing health and social care in a largely rural swathe of north Gloucestershire with 72,000 inhabitants scattered across 49 parishes including the small town of Tewkesbury (Moseley, 1996a). The health authority's strategic planning document had noted that 'distances from the major centres of population and limited transport availability provide a particular challenge in rural areas if we are to ensure that everyone has an equal opportunity to achieve good health' (Gloucestershire Health, 1994: 7).

The author was asked to establish the transport needs and behaviour of four particularly vulnerable groups with typically low levels of car ownership – the frail elderly, adult disabled people, people with learning disabilities and the carers of all of those groups. Of particular concern was access to such services as doctors' surgeries, hospitals, day centres and adult opportunity centres, but also to social and leisure activities, to shops and personal services and, where relevant, to work and training. To that end, surveys were undertaken of the caring agencies, the various transport providers and co-ordinators, the managers of such 'service outlets' as surgeries and day centres, a sample of clients and carers and the area's parish clerks.

In the present context the important point is that the research took a holistic, area-wide view. It looked in the round at the various transport needs of these vulnerable people, at all the transport providers who were relevant to them and at service outlets, as well as the transport that delivered people to them. Scheduled commercial bus services were found to be largely irrelevant to most of the target population because of their infrequency and lack of door-to-door convenience, and because of the physical challenges that buses pose for the infirm.

Table 8.1 sets out as near a complete picture as possible of the minibuses, taxi services, non-emergency ambulances and voluntary car schemes serving those people in that area. It summarises a picture of great variety, complexity and local inventiveness in which the public, private and voluntary sectors all played a significant part. Of the schemes or services identified, 13 used volunteer drivers and 10 used paid drivers. About 32 minibuses or similar vehicles were deployed in one guise or another to meet the needs of the elderly, disabled or frail, and in all about 200 volunteer drivers of cars and minibuses were involved in some formal or semi-formal way.

But provision was geographically patchy and had essentially evolved from a mass of one-off decisions taken by a variety of people or agencies, rather than in a strategically planned or co-ordinated way. Superimposing a series of maps devoted to each type of transport revealed the better served and poorly served areas. Among the former were the parishes close to Tewkesbury or to Gloucester and Cheltenham which lie just to the south of the area. Among the latter was the far east of the district, especially those parishes which lacked a parish-based voluntary car service.

A key conclusion was that the co-ordination of the various transport services was quite minimal. There was certainly little evidence of the various providers sharing the three key resources, namely vehicles, volunteer drivers and information on the complex mosaic of transport need and supply. And the Borough Council's concessionary fare scheme stood completely apart from social services transport provision and from county council public transport co-ordination. Moreover, several minibuses, effectively reserved for just one client group, were significantly underused.

More generally, the 'accessibility management issue' in areas such as Tewkesbury Borough emerged as the challenge of reconciling two desirable but conflicting objectives:

**Table 8.1** ***Transport provision for elderly and disabled people living in Tewkesbury Borough in 1995 – excluding conventional public transport and the person's/household's own private car (from Moseley, 1996a)***

| Service provider | Clients/passengers | Vehicles | The service |
|---|---|---|---|
| **County Council** Tewkesbury Adult Opportunity Centre (SSD) | 60 adults with learning disabilities | Four adapted minibuses | To/from home covering wide area. Five days per week. Also trips during day. Full-time, paid drivers. |
| **County Council** Day centres in Tewkesbury and Winchcombe (SSD) | Approx. 40 elderly people | Two adapted minibuses | To/from home – roughly five-mile radii, seven days a week (Tewkesbury); three days a week (Winchcombe). Also trips during the day. Paid driver (Tewkesbury). Volunteers (Winchcombe). |
| **'Quasi-commercial sector'** Glos Ambulance NHS Trust | Patients referred by GPs or hospitals. Main client was East Glos NHS Trust | One ambulance for most of Borough. Some service also from Cheltenham and Gloucester ambulances | Non-emergency ambulance service to take in-patients or out-patients to hospitals within or outside area. In a competitive situation: GPs and hospitals could use taxis, voluntary car schemes etc. instead. |
| **Commercial sector** Two taxi operators | General public but with specific provision for disabled people | At least two taxis adapted for wheelchairs | Only about two thirds of the area was effectively covered by adapted taxis. |

*(Continued)*

**TABLE 8.1 *Continued***

| Service provider | Clients/passengers | Vehicles | The service |
|---|---|---|---|
| **Commercial sector** Minibus hire firm | Children with special needs and disabled people | Five minibuses taking wheelchairs | Mainly special needs children to school. But private bookings from whoever. |
| **Voluntary sector** 'Dial a Ride': three services run by voluntary bodies based outside the borough in Cheltenham, Gloucester and Newent | Frail elderly people and disabled people | In all, 16 minibuses with tail lifts. Parts of the Tewkesbury Borough were not served. | Trips to hospital, GP, dentist, day centres, shops, hairdressers etc. Typical cost to passenger of £2–3. Grants from County Council. |
| **Voluntary sector** *Minibuses* (i) an association of five churches (ii) wheelchair bus association (iii) 'old people's welfare committee' | (i) any group in the parishes (ii) elderly and physically disabled (iii) elderly people only | Five minibuses of which four were adapted. | Only (iii) operated across whole borough; others were localised. Volunteer drivers often in short supply. |
| **Borough-wide voluntary car schemes** (i) Tewkesbury volunteer help service (ii) WRVS voluntary car service | Mainly elderly people (i.e. problems of getting in/out of car for disabled people) | Private cars: each scheme had a pool of approx 20 drivers. | Served all/most of area. Typically to hospitals, medical appointments, some shopping and social. Weekends/evenings not well served. County Council support. Typical cost to passengers about £3. |
| **Five very local voluntary car schemes** (*Note*: Glos Ambulance Service and Tewkesbury Day Centre also had own volunteer car drivers) | Varied. All specified elderly/disabled. Some added 'those in transport need' | About 70 drivers involved in the borough. | More localised, but otherwise as above. Only about half of Borough covered by these local schemes. |

- achieving the scale economies and reduction of excess capacity that might be expected to arise from co-ordination by a central unit; and
- releasing the local entrepreneurship and 'common sense' that comes from a highly decentralised system of resource management.

Our conclusion was that, on balance, it would be desirable to move towards a greater measure of co-ordination, incorporating the notion of 'brokerage'. This would involve equipping someone or a small unit to influence the provision of each specific service – its route, timing and clientele – according to the pattern and nature of transport needs as they evolved from day to day. Success would depend partly on that person or unit enjoying excellent communications with those needing door-to-door transport and those able to provide it, and partly on the development of an area-wide culture of sharing and collaboration rather than of 'turf protection'. Happily, the ensuing years saw some significant steps taken in that direction.

---

## CASE STUDY 16: THE WEST NORFOLK COMMUNITY TRANSPORT PROJECT

---

In west Norfolk a brave attempt was made in the mid-1990s to grapple with issues of co-ordination very similar to those described above in relation to the Tewkesbury area. This initiative was researched and written up in 1997 by the present author as an example of an 'area-wide partnership' – an arrangement which 'involves two or more agencies with a mutual interest in the delivery of a single service working together to provide it across a given geographical area' (Moseley and Parker, 1998: 32).

In 1991 the Social Services Department of Norfolk County Council resolved to respond to the chronic insufficiency of transport available to disabled and elderly people living in the sparsely populated rural west of the county. It convened a West Norfolk Transport Liaison Group, involving the transport development officer of the district council (the borough of King's Lynn and West Norfolk) and other agencies in the health and social services fields, as well as the King's Lynn Volunteer Bureau. A project founded on the good sense of better use of underused vehicles for the common good was thereby born.

In 1992, thanks to funding provided by the county and district councils, the district health authority, the county's family health services authority, Help the Aged and the Rural Development Commission, a co-ordinator was appointed and installed in a small office in King's Lynn, with an assistant co-ordinator joining him a year later. A computer system was installed to help the planning of routes and services and to enable the co-ordinator to broker the matching of the supply and demand for transport across the area. In 1994 the project took over three minibuses, each with their own driver operated by the Volunteer Bureau, and there was a steady increase in the range of services offered and the use made of them over the next three years.

The service as it developed, therefore, comprised an integrated approach to the delivery of transport for elderly and disabled people across an extensive local authority district. The co-ordinator, working to a committee drawn from a range of statutory and voluntary bodies and with user representation, essentially sought to meet demand by maximising the use of available seats in a host of different vehicles.

The vehicles committed to the scheme (Rural Development Commission, 1996a) were owned by the social services department (16 vehicles including six adapted to meet the needs of people with physical disabilities), the health authority and charities such as Age Concern, but also included were volunteer car schemes and, where necessary, taxis. The project itself owned eight vehicles which were effectively added to the pool. A whole host of services was thereby provided including a ring-and-ride service, a community care service, a group hire scheme and a primary healthcare service which transported clients from their homes to their doctor's surgery.

The result was that the elderly and disabled people of the district's rural areas, who comprised three quarters of the service's clientele, got a transport service better able to meet their specific needs and a better quality of life in consequence. As for the transport providers, they had access to a wider pool of vehicles for their clients and, if they wished it, first-line management of their drivers. In addition, the voluntary groups that participated in the scheme got help with the maintenance of their vehicles, help with legal and safety issues, help with finding volunteer drivers and an income from making their vehicles available to other users. In effect they could benefit from certain scale economies which derived from the overall operation.

Set against these benefits were the operating costs of the scheme (some £120,000 in 1997), met in roughly equal measure by a levy on fares and a subvention from the partner agencies, and the loss to each agency of the flexibility of having its own vehicles sitting idle awaiting use.

Looking at the scheme as an example of 'innovation', factors favouring the innovation included:

- a large area with a scattered population creating very similar problems for several agencies;
- a growing realisation by each agency that 'go it alone' was not cost-effective;
- the existence in King's Lynn of both a transport development officer, employed by the district council, and an active volunteer bureau; and
- one agency – the social services department – prepared to 'grasp the nettle' and put together a planning group and a funding package.

Constraining factors had been:

- an initial reluctance by voluntary bodies owning hard-won vehicles to share them, especially if they were the product of a long process of fundraising; and
- a history of transport being treated as an afterthought in health and social service planning. In the early days people would come to the project manager and say 'We have provided a new day centre in such and such a place; now what can we do about transport?'

Clearly the scheme went some considerable way towards achieving the 'holistic approach to accessibility management' advocated in this chapter, though the reader will be aware of how much remained outside this particular 'umbrella'. The lessons that might be distilled as good practice include the need to ensure:

- the early establishment of a widely drawn steering group involving the statutory and voluntary sectors;
- a good management committee representing the various stakeholders;
- good links with a volunteer bureau to ensure a supply of volunteer drivers (often a key constraint) and access to the voluntary sector groups that might wish to use the service;
- ensuring the provision of quality and reliable service to gain and retain confidence;
- encouraging an entrepreneurial, innovative culture in which new ideas and developments are constantly being taken on board; and
- carefully reconciling a need to 'balance the books', meaning a quasi-commercial remit, with the 'community transport ethos' necessary to retain legitimacy with the various stakeholders.

---

***SEE ALSO...***

Other case studies which concern, in part, the provision of transport and other services at the local level include numbers 3, 6, 11 and 13.

***SELECTED FURTHER READING***

Moseley (1996a and 2000c), National Rural Enterprise Centre (1998) and LEADER Observatory (1999e) collectively cover many aspects of rural transport, service provision and the potential of telecommunications in that domain.

# 9

# Partnership: Working in Harness

> 'I thought,' said Piglet earnestly, 'that if Eeyore stood at the bottom of the tree, and if Pooh stood on Eeyore's back, and if I stood on Pooh's shoulders...'
>
> A.A. Milne, *The House at Pooh Corner*

## DEFINITIONS AND IMPORTANCE

A recent research report on rural regeneration partnerships in mid-Wales and Shropshire (Edwards *et al.*, 2000: 1) observed that

> partnerships have emerged over the course of the past 15 years as an increasingly commonplace fixture on the landscape of urban and rural regeneration in Britain ... partnership working was developed in the 1980s by the British government as a means of loosening local government's hold over urban economic development and by the European Union as a mechanism for redirecting its Structural Funds from large-scale state-led projects to small-scale bottom-up initiatives [but] the 'partnership principle' has now become established as a preferential mode of management across a diverse raft of policy arenas.

To reinforce their point, the authors went on to note that

> close examination of the organisations operating in any small town in England or Wales is likely to reveal LEADER groups, Local Agenda 21 groups, training partnerships, community enterprise or development projects, civic fora, and rural development programmes as well as a plethora of groups focused on marketing, product valorisation, sustainable development, transport or tourism – all constituted as some form of partnership, bringing together a range of organisations often from across the public, private and voluntary sectors. (*ibid.*: 1)

This observation was reinforced by their finding as many as 150 partnerships in some way involved in 'rural regeneration' in just three counties, Ceredigion, Powys and Shropshire, in 1998–99.

A recent and important example of the creation of local partnerships is provided by the government's determination, expressed in the Local Government Act 2000, to have 'community strategies' created and implemented at the local authority level in England and Wales. These strategies are intended to 'improve the economic, social and environmental well-being of each area and its inhabitants and to contribute to the achievement of sustainable development in the UK' (Department of the Environment, Transport and the Regions, 2001: 2). To that end 'local strategic partnerships' have been set up involving public, private community and voluntary sector organisations in the hope that they will bring together, and in some circumstances replace, many of the other partnerships and plans alluded to in the previous paragraph. Certainly the wish is that these 'community strategies' will provide a well-researched overarching context for the preparation and implementation of other more specific plans and programmes serving the locality in question.

The present chapter focuses on this proliferation of local partnerships – especially those explicitly concerned with area-based rural development – and looks particularly at their operation and their alleged ability 'to add value' to the development process in ways that similarly resourced *individual* agencies would find difficult or impossible.

There is by now a considerable literature on this phenomenon (see, for example, Hambleton *et al.*, 1995; OECD, 1996; Slee and Snowdon, 1997; LEADER Observatory, 1997d). But here we will draw particularly on the cross-national PRIDE research project in which the authors of this chapter were involved between 1999 and 2001 – PRIDE standing for Partnerships for Rural Integrated Development in Europe. That study spanned six countries (Finland, Germany, Italy, Spain, Sweden and the UK) and researched 330 local rural development partnerships, 24 of them in considerable detail (Westholm *et al.*, 1999; Esparcia *et al.*, 2000; Cavazzani and Moseley, 2001; Cherrett and Moseley, 2001; Moseley, 2001; Moseley, 2003).

Edwards *et al.* (2000: 2) define a *partnership* as 'an arrangement which deliberately draws together the resources of specified partners in order to create a capacity to act with regard to a defined objective or set of objectives', stressing the importance of *deliberate* creation and the assembly of *resources* (not necessarily financial) with predefined *objectives* in mind. Focusing more specifically on rural development, the PRIDE project produced the following formulation: 'a voluntary alliance of organisations from at least two societal sectors (state or public sector organisations, private companies, civil associations) with a clear organisational structure, with ongoing and long-term activities that include more than one project and which show an integrated approach to the promotion of development of rural areas with no more than 100,000 inhabitants'. (Moseley, 2001: 6) – not so much an elegant 'definition' as a set of guiding criteria agreed by the PRIDE researchers to ensure that the six national teams focused their work on broadly similar phenomena.

This rise to prominence of partnerships as the preferred machinery for promoting local rural development is one expression of a fundamental shift in the way that local areas are governed and serviced, a shift away from the notion of govern*ment*, with the state and the elected local authorities unequivocally centre-stage, to one of govern*ance* which acknowledges the inter-related roles played in local areas by a host of actors including the state and the local authorities, but also embracing private businesses, voluntary organisations and community groups, and also supra-national institutions such as the European Union. This loosening up of 'government' to give 'governance' has a complex set of causes. Not least among these have been some key changes in the culture of many European national governments in recent years to embrace a greater decentralisation of decision-making, deregulation and the creation of an 'enabling role' for local authorities which requires them increasingly to contract with private firms and voluntary sector organisations to deliver services and promote development on the ground while they, the local authorities, retain ultimate responsibility.

It follows that a definition of *governance* may also be helpful, such as that put forward by Stoker (1996) and reproduced in Goodwin's review (1998: 5) of the governance of rural areas: 'governance refers to the development of governing styles in which boundaries between and within public and private sectors become blurred'. Or, linking this blurring to the rise of partnership working, 'governance is about governmental and non-governmental organisations working together in non-hierarchical and flexible alliances' (Murdoch and Abram, 1998: 41). What we are concerned with in this chapter, then, is the role and effectiveness of such 'non-hierarchical and flexible alliances' in the promotion of local rural development.

## SOME KEY ISSUES

The international literature review (Westholm *et al.*, 1999) that formed a launch-pad for the PRIDE research project highlighted several widely expressed concerns about the growing involvement of partnerships in local development and the delivery of services to rural people:

- Many local partnerships are essentially a 'tactical response' to the need to secure funding from outside sources such as the European Union and national governments. In consequence, they may not be truly grounded in the local community and its needs and may not survive the expiry of the particular funding stream.
- Local partnerships are often dominated by the public sector, especially the local authorities and local and regional agencies of central government. Genuine local community involvement is frequently weak, with community and private sector representatives often unclear about their roles.
- When there is a genuine delegation of responsibility, questions of accountability arise, with unelected private and voluntary sector partners substantially involved in what are really political decisions about resource allocation.

- Because of their need to proceed by consensus, partnerships can often be rather conservative bodies supporting safe rather than adventurous initiatives. This can be particularly true if the partnership has effectively been put together by 'the establishment' with its tendency to favour organisations that are traditionally well represented in the area – what one commentator called 'rounding up the usual suspects'.
- In practice, most partnerships are much better at delivering discrete projects, albeit in a variety of sectors, than truly integrated programmes.
- Local partnerships may widen rather than narrow social and economic disparities if those which are successful tend to be rewarded with further funding, and vice versa.

The subsequent research tended to confirm those concerns to often have some validity, while also finding excellent examples of partnerships genuinely 'adding value' to the local development process. But it is hard to generalise since, as Edwards *et al.* (2000: 10) noted in their mid-Wales/Shropshire study, 'the growth of partnership working in rural regeneration has not produced a new homogenised form of rural governance institution – 'the partnership' – but rather a diverse and complicated menagerie of partnership organisations with different foci, different scales of operation, different durations and histories and different patterns of sector representation and funding'.

Turning now to the UK component of the PRIDE research, a survey of 54 local partnerships, all respecting the working definition set out earlier, confirmed this variety (Cherrett, 2000). That sample, to some degree inevitably reflecting the vagaries of responses to a postal survey, comprised 13 LEADER groups, eight English Rural Development Programmes, eight boards or committees of National Parks or Areas of Outstanding Natural Beauty, and seven Market Town Regeneration partnerships, plus a number of Local Agenda 21 initiatives, various agriculture or land management programmes, Rural Challenge partnerships and others which were hard to classify.

Focusing on those 54 rural partnerships,

- the most common reasons given for their *formation* were to access funding, to pool expertise or other resources so as to address common problems and to provide a vehicle for community involvement. There were very few genuinely 'bottom-up' partnerships in the sense that the real thrust had come 'from below' rather than from existing powerful players;
- their *members* were drawn mainly from public agencies, local government, the voluntary sector and local community groups – the private sector tending to be poorly represented. Roughly equal numbers of the partnerships had a formal status, often that of charitable limited companies, or else an informal status for which some sort of 'memorandum of agreement' sufficed;
- as for their *objectives*, most respondents in the survey made some reference to 'integrated' and/or 'sustainable' development, though in practice one or

more of the following tended to prevail – economic regeneration, community involvement, the promotion of tourism/recreation, and environmental conservation;

- the amount of genuine *community involvement* that the partnerships practised varied. Some went no further than involving a small number of community organisations on their boards and others construed 'community involvement' as merely ensuring that many recipients of project funding were in some guise 'the local community'. But others did employ more formal mechanisms of consultation and involvement such as village appraisals and 'planning for real' exercises;
- finally, asked about their *outputs*, most partnerships made rather more of their 'soft outputs' such as 'mobilising the community', 'raising awareness and understanding' and 'raising the area's profile' than they did of 'hard outputs' such as 'business start-ups' or 'jobs created'. Whether this coyness regarding the latter concealed some disappointment with their more tangible achievements, the fact that it was still 'early days' or an appreciation that such output measures are fraught with difficulties (who actually *creates* businesses or jobs?) was not clear.

Those local partnerships – very much the majority – which did report some measure of success on the ground tended to attribute it to one or more of the following factors. First and foremost was the availability of funding and associated success in 'levering' in further, local funding. Then, in no particular order, came the active support of the local authority/ies and the funding partner(s); the active support of the local community as a source of local knowledge, energy and labour; a good co-operative atmosphere and real collaboration between the partners; and the availability of high-quality staff and/or individual board members prepared to run with certain issues or areas of work.

Asked about the main *constraints* that inhibited their work, the lack of those five success factors, where applicable, clearly had its effect. But the most frequently mentioned constraint was 'excessive bureaucracy' and an associated lack of real devolution of authority to the partnership. Of particular concern were the time-consuming and 'belt and braces' procedures which had to be negotiated both before getting formal project approval from the many funding agencies and also in subsequently drawing down funding from national and, more especially, European sources. Procedures that might be justifiable where a few very large projects are concerned are much less so in relation to dozens of small projects each typically costing just a few thousands of pounds and involving relatively unsophisticated community groups.

**But do partnerships, *per se*, really 'add value' to the local development process?** This is the key question; if the answer is 'no', then individual agencies such as the local authorities may as well be given the duties and the resources assigned to partnerships and allowed to enjoy all the advantages of going it alone. In fact, the answer is 'typically yes'. At the end of the whole PRIDE research project, and pulling together the evidence from 24 extended case studies in the six countries which together involved several hundred interviews spanning both partnership operation and partnership impact, a simple model emerged to summarise the genuine 'partnership' contribution to the delivery of projects and programmes on the ground (see Figure 9.1).

In essence, local partnerships contribute two crucial intrinsic characteristics (row A) to local development which individual state agencies acting alone simply cannot provide; they are the *bringing together* of a variety of actors and *quasi-independence* from state control and procedures. These two characteristics underlie a number of *potential capabilities* (row B) which, properly exploited, can and do really add value to the development process. But the absolutely central point is that these capabilities cannot be taken for granted; they do not just 'happen' as soon as a partnership is created. They need to be *unlocked* by systematic attention being paid to *four key elements* (row C) which relate to the partnership's individual partners and to its resources, organisation and processes. The research showed that doggedly working at those four elements (row C) so as to unlock those capabilities (row B) produced a rich harvest in terms (row D) of sustainable, endogenous, innovative *development on the ground*. The metaphor that seemed most apt was that of turning the 'keys' (C) to open the 'doors' (B) to allow real partnership-based development impulses (D) to flow into the area in question.

## THE TOOLKIT

Our toolkit for increasing the effectiveness of local rural development partnerships simply builds on the significance attributed to the 'partners, resources, organisation and processes of partnerships' in Figure 9.1, so as to create a hypothetical *model local partnership* devoted to the socio-economic regeneration of some small rural area. Obviously, as local circumstances and partnerships themselves both vary so much, there can be no single 'model' as such. But this one, based on the lessons of the PRIDE research project and distilled in a Good Practice Guide (Cherrett and Moseley, 2001) has been explicitly designed to help both the sponsors of local partnerships and the partnerships themselves to reflect constructively on their experience. It draws extensively on the experience of the four UK partnerships that were studied in depth, two of them summarised as Case Studies 17 and 18. The argument is that by respecting the 20 guidelines listed below, more effective partnership working will ensue.

So, necessary for a 'model partnership' intent on promoting local rural development are the following:

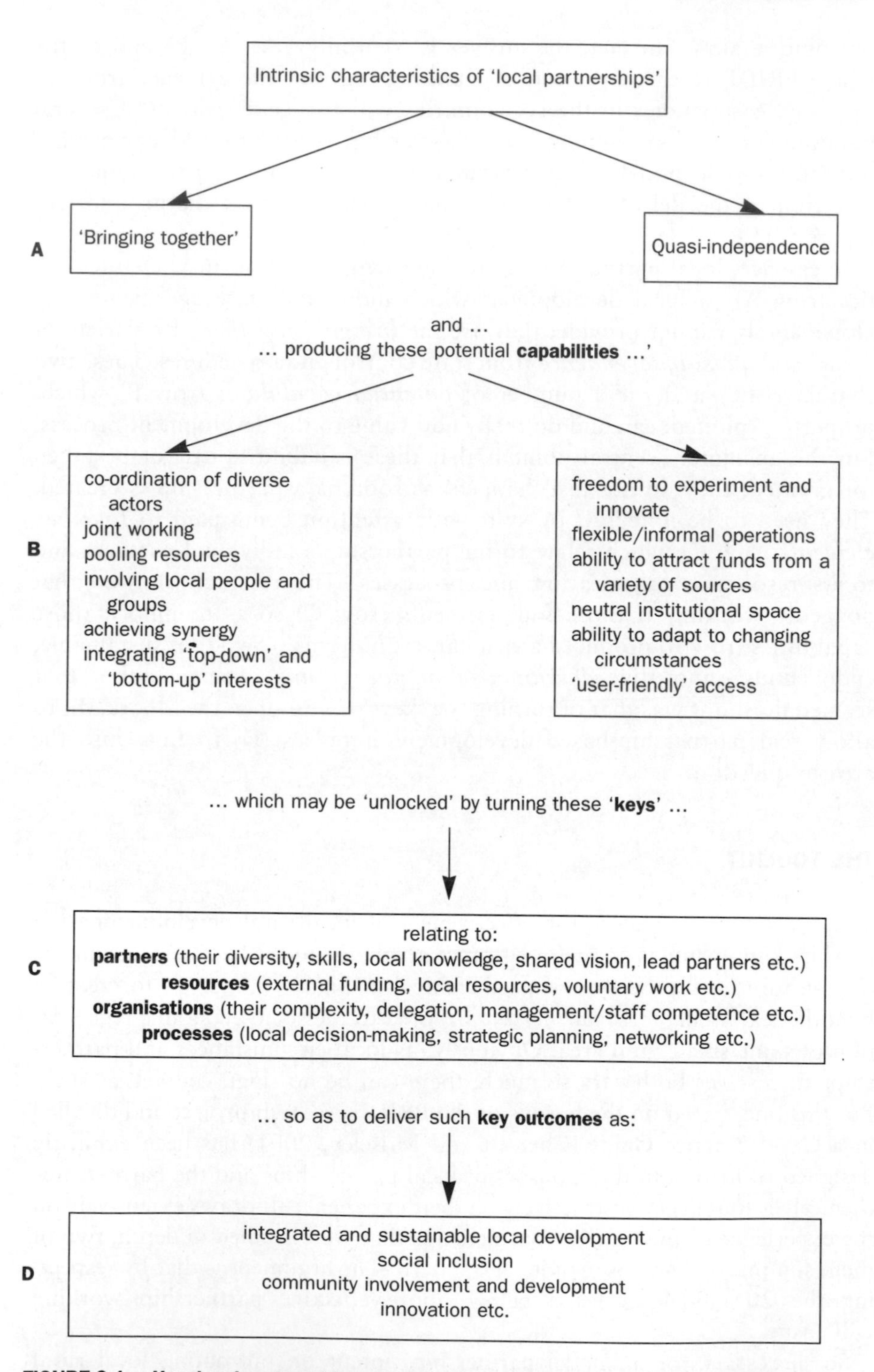


**FIGURE 9.1** ***How local partnerships add value to the development process (Cherrett's model, reported in Moseley, 2001, 2003)***

### THE INITIATION OF THE PARTNERSHIP

1. One or more 'lead partners' consults with a wide range of other public, private and voluntary organisations and local community groups to identify key local issues that need to be tackled 'in the round'.
2. A broad, but draft, statement is produced describing the vision and goals of a collaborative venture to address those issues, and highlighting what resources might be necessary.
3. The lead partner convenes an open meeting to discuss the draft proposals and a working group is set up to develop the proposals in more detail.
4. The working group then produces a more detailed proposal spelling out the venture's aims, vision, objectives, method, resources and management. Potential partners are invited to discuss and 'sign up'.
5. Widened to include new members if appropriate, the working group then draws up a detailed bid for funding drawing on the expertise and local knowledge of the various partners and respecting what contributions they would be able to make.
6. A funding bid, or bids, is made, and if successful, or if the partnership decides to go ahead anyway, the working group draws up detailed 'terms of engagement' indicating the partners' commitments, to which they sign up. A constitution is agreed, a chair and management committee are appointed and the work begins in earnest.
7. The partnership is launched and widely publicised and a door is left open for others to join in.

### ITS PLANNING AND MANAGEMENT

8. Based on the seven points above, a clear operational 'business plan' is produced, linking agreed objectives and targets to the projects to be pursued and to the resources – financial, material and human – needed to achieve them. The plan also sets out the wider context of the partnership's work including the main links to be made with other agencies and programmes serving the area.
9. The partnership resolves to revisit and if necessary revise its business plan in the light of changing circumstances, and it agrees when it will have to prepare a 'roll-forward strategy' or else an 'exit strategy'.
10. Clear management structures and procedures are established relating to overall control (a board?), day-to-day management (an executive committee?) and particular matters such as grant applications (sub-committees or groups with some measure of open membership?).
11. First-class staff support is recruited, to include a project manager or co-ordinator with commitment, energy, flexibility and skills in networking and administration.
12. Steps are taken to ensure that decision-making processes are open, accountable and accessible to members and to the wider community, and to encourage ever wider involvement.

### *IMPLEMENTING THE WORK OF THE PARTNERSHIP*

In carrying out its work, a model local development partnership

13. is 'user friendly' to its existing and potential partners through clear management processes, good information provision and a culture of open access and involvement;
14. is 'user friendly' to existing and potential recipients of its support by providing easily accessible information, advice and support, by using easy-to-use funding procedures and by ensuring good follow-up once projects have been launched;
15. is accessible to minority and hard-to-reach groups in the community, developing proactive ways of reaching, encouraging and involving them;
16. exploits its relative independence from the statutory agencies by being both innovative and flexible; and
17. encourages partners to collaborate closely and so generate the synergy that should be the bonus of partnership working; the whole should be demonstrably greater than the sum of the parts.

### *LOOKING TO THE FUTURE,*

a model partnership

18. provides support and training to raise the capacity and confidence of local people and organisations;
19. develops a longer-term programme of work to be planned and managed by the partnership itself or by a successor body or bodies; and
20. celebrates its successes and achievements in ways that encourage greater local ownership and involvement.

The two case studies that follow incorporate many of these commendable features. They were researched by the authors of this chapter as part of the PRIDE project referred to above (see Cavazzani and Moseley, 2001 for fuller reports on each). The italicised quotations all come from local interviewees in those studies.

## CASE STUDY 17: THE SOUTH-WEST SHROPSHIRE 'RURAL CHALLENGE' PARTNERSHIP

---

In 1994, England's Rural Development Commission announced its Rural Challenge programme which was designed to encourage truly enterprising local responses to the socio-economic problems of rural England. The programme committed the Commission to spend £1 million over three years in each of a few select localities chosen on the basis of detailed bids put forward by putative local partnerships able to harness local energy and a good measure of co-finance. The successful south-west Shropshire bid, masterminded by the chief executive of the South Shropshire District Council, related to an area of 24 parishes close to the Welsh border, centred on the small town of Bishop's Castle and containing about 10,000 people spread over 345 sq. km. It was an area that had recently lost two major employers, had unemployment running at 12 per cent and an air of decline, as evidenced by boarded-up shops in the town.

In interview, the district council's chief executive explained how the partnership of 18 member organisations was put together: '*We started with the district council; obviously we also had to have the county and town councils, the town's community college had a strong record in community development and there were two firms ready to invest heavily in local projects...*' Thereafter invitations went out and, as someone observed, '*with a million pounds to spend it seemed a good idea to get involved*'. The eventual partnership of 18 organisations (plus an independent chairman – a respected industrialist from outside the area) comprised nine truly local partners including Bishop Castle's Chamber of Trade and various local forums of parish councils and community groups, plus nine district or countywide organisations including the district and county councils, the Training and Enterprise Agency, the Community Council of Shropshire, the health authority etc.

The goal was '*to follow a model of development spanning the social, economic and environmental dimensions which we knew could not be looked at in isolation*'. Consequently, the successful Rural Challenge bid comprised 17 proposed social and economic projects, which those partnership members not involved from the very start had to accept as something of a *fait accompli*, though, as circumstances changed and some of the agreed projects proved abortive, the partners grew to feel that they '*had something to play for*'.

The organisation and operation of the partnership may be summarised as follows:

- The 'Lifeline Company', limited by guarantee and without share capital, was created to give the partnership a legal status, with the district council's chief executive as its company secretary and, *de facto*, its main driving force.
- Every three months there were rather unwieldy board meetings comprising the chairman, 18 partner representatives and often, in addition, their deputies and various other 'observers' from the partner agencies.
- A salaried project manager was charged with the day-to-day facilitation of the whole enterprise, liaison with the various interest groups, project promoters etc.

- Six highly successful sub-groups or panels were created, each with 5 to 15 members, by no means all of them 'formal partners'. These groups '*brought more local people in and gave them a sense of power*'. For example, one group managed 'Enterprise House' which was a redundant factory in the centre of Bishop's Castle converted into a resource centre with business units; another group devised and promoted tourism initiatives; others managed small grants funds which were earmarked for bottom-up community ventures or else for small-business initiatives.
- Frequent liaison was maintained with the regional officer of the Rural Development Commission who remained supportive and constructive throughout but whose job of ratifying and, if necessary, renegotiating decisions reached by the partnership on project funding sometimes caused delay and some resentment.

There were inevitably disagreements to deal with, a notable one relating to the Community College's disputed suitability to house the IT training facility and associated resource centre. And some initial mistakes were made; for example more might have been done in the early months to inform and involve the local community who mistakenly thought '*we've won £1 million, what shall we spend it on*?' when they read in the press of the success of the project application, and who then became disillusioned when little was visibly happening on the ground in the first year.

But, by any reasonable standards, the three-year Rural Challenge programme was a great success. Comments made to the author early in 2000 included: '*Yes, it's delivered on the economy – look at Enterprise House and Challenge Court*' (a complex of small business units all in occupation); '*Yes, the Lifeline Company and the College are still there*' – despite their disagreements – with the company being 'rolled forward' to keep up the momentum after the money ran out in 1998, and the College, with improved facilities, remaining a significant player in the local community. Further observations were: '*Yes, it has brought the different interests and the community together*' ... '*Have a look round...you won't see any empty shops now*' and '*Before 1995 people round here were pretty good at protesting; now they have a belief that they can do something for themselves*'.

In financial terms, in just three years the partnership apparently brought £5.2 million into this area of only 10, 000 people – 'apparently', because it is difficult to estimate what would have happened without the Rural Challenge programme. Be that as it may, the £5.2 million comprised the Rural Development Commission's £1m, plus substantial co-funding of the chosen projects from the private sector and the district council, and also lesser sums from other arms of central government, other local authorities and the European Union.

**Why this success?** The PRIDE research suggested a number of key factors over and above the availability of a significant sum of money to spend. They included:

1. the drive and leadership of the district council and, in particular, its chief executive;
2. the real commitment and constructive involvement of certain other public sector partners, especially the Rural Development Commission, which was much more than just a source of cash;
3. genuine community involvement building up over time – *'Scratch a town and all sorts of people come forward with talent and experience and once they feel engaged, things develop'*;
4. a talented and energetic project officer in the key middle year of the programme (there were problems before and after that year);
5. good chairmanship by a respected figure with a distinguished business record and a degree of detachment – *'He could beat heads together...he could stand back and find a way through'*;
6. the right size of area – *'At this scale people give their time freely for the community; any smaller and you lack a critical mass, much bigger and you get the wrong mental attitudes...I certainly wouldn't have done all this in Wolverhampton'* (from where the interviewee originally came);
7. the considerable use made of small 'action groups' which intermingled board members and other local people and enjoyed a good measure of delegated authority;
8. over time, the build up of a good measure of 'local ownership' – achieved without abandonment by the key agencies;
9. the fact that the funding was high-profile, the 'Rural Challenge million' grabbing people's interest; and
10. the fact that some of the *spending* was also high-profile, notably the two hugely popular 'small grant schemes', neither of them disbursing more than £1000 to any individual applicant but each being highly visible and producing tangible 'mini-projects' with which people could easily identify.

But with hindsight, and recalling both the great rush to get the initial bid together and set up the partnership, and the local disillusionment in the first year when nothing seemed to be happening, a four-year programme would have been better than one of three years. *'Everything needs time ... we needed a soft entry to build up community ownership of the programme and to get people on board.'*

---

## CASE STUDY 18: THE WEST TYRONE RURAL 2000 LEADER PARTNERSHIP

---

West Tyrone Rural 2000 was one of 15 LEADER II partnerships in Northern Ireland. It is of particular interest here because, in an area riven by political strife and sectarian division, it successfully brought together people from both the Unionist and Republican communities to work in common cause to promote the economic vitality and prosperity of the area. LEADER

II funding from the European Union amounted to some £740,000 and this was supplemented by substantial co-funding from national and local sources.

West Tyrone, in the far west of Northern Ireland, comprises the two local authority districts of Omagh and Strabane and has a population of some 82,000 spread over 2000 sq. km. It has suffered for many years from the combined effects of geographical peripherality, a weak economy based largely on agriculture, high unemployment (running at 17 per cent when LEADER II was launched), material deprivation confirmed by several indicators and a lack of confidence arising from decades of harsh political conflict and terrorist activity.

Early in 1996 the two district councils successfully made a joint bid for LEADER II status and launched West Tyrone Rural 2000. Thus, as in the Shropshire example, it was a combination of a top-down funding opportunity and a coherent bottom-up response from the local authorities that underlay the birth of the partnership. The award of project grants then proceeded until the end of 1999 with a further year or so to complete actual expenditure, undertake some evaluation and consider any future action.

The main partners were:

- *The local authorities* – Omagh and Strabane district councils;
- *Government offices and agencies* – most notably Northern Ireland's Department of Agriculture (Rural Development Division), the Local Enterprise Development Unit (LEDU) and the Western Education and Library Board;
- *Farming unions and associations* – the Northern Ireland Agricultural Producers Association and the Ulster Farmers Union;
- *Community representatives* – Strabane Community Network and the Omagh Forum of Rural Associations; and
- *The Private sector* – Omagh Tourism Group and small and medium enterprise representatives.

Interviews with the partners revealed that they valued the chance to play a responsible role in local affairs and to be able to converse regularly with other players whom they might not normally meet. One local councillor remarked, '*At first I thought it* (the partnership) *might be a bit of a 'junket' but I found it to be very interesting and worthwhile. The projects are very helpful for local business and there has been a strong success rate. It was and is much needed in a very peripheral area such as this, with economic decline and a lack of confidence.*'

The partnership worked to a set of goals espousing all the familiar elements of multi-sectoral, integrated rural development, but in practice it put the emphasis firmly on economic development, recognising the need to diversify a largely agricultural economy. As the partnership's project officer put it, '*There was an economic focus from the start; we wanted to "head for the private sector". There were other funding programmes available for community projects but, apart from LEDU, there were none for private business*'.

The partnership was structured around a main board meeting held roughly monthly, and five sub-committees, one of them effectively an executive committee handling day-to-day business and general management issues and four 'sectoral' sub-committees responsible for delivering the board's strategy and assessing project proposals on a thematic basis. Staff support was provided by a project manager and an administrative assistant. One partner remarked '*One of the main strengths of the partnership is that we have a good project officer with excellent experience and knowledge of the area.*'

In essence, the work of the partnership and its sub-committees was to solicit, appraise and, if appropriate, allocate funds to projects meeting the LEADER II guidelines and the objectives set out in the partnership's own development strategy. To this end the work was mainly *re*-active though some innovative *pro*-active initiatives were also taken by the Board including developing, and then playing an active role in, the Tyrone/Donegal Cross-Border Partnership.

An important feature of the partnership's operation was the hands-on involvement of many of the directors (the individual people who represented the partner organisations on the board), despite the fact that '*the officers did their homework well and we, the directors, were always well-briefed for decisions*'. As well as sitting on sub-committees, individual directors got involved in publicity and promotion, developing proactive partnership projects and networking, often in addition taking a particular interest in specific projects.

Despite a lack of interest from certain quarters, notably older farmers with small farms who were '*too used to "farming the grants"; they find it difficult to break into new ways of working or new initiatives*', the partnership identified more eligible projects than it had funds available. A wide range of initiatives was supported, including the Irish Border Food Directory, the promotion of traditional stone working, a village enhancement scheme, a 'fun farm' and a stone-crushing plant. No project got over £50,000 and the average grant was substantially less than that.

To conclude, we may note that, having examined West Tyrone Rural 2000 in some detail, the PRIDE researcher, Cherrett (2001), was persuaded that some aspects of its operation could usefully be re-examined and, if necessary, strengthened. These included the accountability to the local community of this unelected body, the rigour of the linkage between the partnership's strategy and the real needs of the local area, the quality of the collaboration between the partnership and the many different agencies, programmes and other partnerships with related concerns and objectives, and the composition of the partnership, given that it contained very few women or younger people.

But the partnership had clearly proved its worth in stimulating business enterprise in the area. In this respect its success factors were similar to those we have seen in the Shropshire example. These included the positive involvement of both the local authorities and the relevant government agency, genuine community involvement, good project officer and administrative support, a clear sense of purpose linked to an agreed development strategy and, of course, substantial external funding which was enough to 'make a difference' but not enough to obviate the need to persuade local sponsors to get out their chequebooks too.

But none of this would have proved effective if the whole venture had not sufficiently captured the imagination of the partner organisations and of the individual people who represented them to persuade them to leave their 'political baggage' outside the committee room. In the Northern Ireland context of the late 1990s this was a remarkable achievement and all the partners interviewed emphasised this, one observing that '*Everyone is supportive, the partnership gels; you can't afford to bring in a private agenda*'. Indeed, looking beyond the particular context of Northern Ireland, there may be a useful suggestion here that establishing a tradition of local partnership working may not only help the process of local socio-economic development but may also serve wider goals relating to social and political cohesion and civic ideals.

## NOTE

This chapter has been written jointly with Trevor Cherrett of the Sussex Rural Community Council.

## SEE ALSO...

Case Studies 2, 9, 14, 16, 20 and 23 also present local partnerships of different sorts.

## SELECTED FURTHER READING

See Goodwin (1998) on local governance, and Edwards *et al.* (2000) and OECD (1996) for national analyses of rural partnership working (Wales and Ireland respectively). See Slee and Snowden (1997) and Cherrett and Moseley (2001) on good practice.

# 10

# Community Involvement: Embracing the People

> Because half a dozen grasshoppers under a fern make a field ring with their importunate chink ... do not imagine that those who make the noise are the only inhabitants of the field.
>
> Edmund Burke

## DEFINITIONS AND IMPORTANCE

The increasing involvement of local people in rural development has been a striking feature of the past few years, whether in relation to the formulation of strategies for action or in the actual delivery of programmes on the ground. Thus the government's 'Rural White Paper' (Department of Environment, Transport and the Regions and MAFF, 2000: 145) stressed that 'we want to see...people living in rural areas being fully involved in developing their community, safe-guarding its valued features and shaping the decisions that affect them'. And local development initiatives resting on the 'challenge principle' as with Rural Challenge and the Single Regeneration Budget and/or on the operation of local partnerships, as with LEADER and England's Rural Development Programmes in the 1990s, have all extolled the benefits of carefully encouraging the active involvement of the people living in the relevant area.

The growing weight placed by European, national and local government on community involvement in rural development reflects its wish to reap four anticipated benefits (Warburton, 1998a; Lowe *et al.*, 1998) as follows:

### *BETTER DECISION-MAKING*

The first argument has two strands. Local people, if carefully consulted, are a source of valuable ideas, information and wisdom that it would be folly to ignore, the resultant programme being better targeted and more cost-effective for having culled local knowledge and opinion. In addition, the burden of adversarial decision may be reduced by seeking at an early stage to reconcile conflicting views. In this connection there is often an 'educative element', that of encouraging local people to see the bigger picture and to appreciate the needs of other groups and of neighbouring areas; this conception of the purpose of community involvement often hopes that it will be an antidote to NIMBYism in its most myopic form.

### *MORE DURABLE ACTION*

Second, to the extent that any proposed policy or action is built upon a genuine local consensus, it is more likely to be durable and to escape being scuppered by local antagonism or indifference. Indeed, if the action has local origins and 'ownership' – if it has emerged from the involvement exercise – then it is more likely to be pursued with vigour and sustained commitment by local people even, perhaps, when the initial financial resources have run out.

### *THE PROMOTION OF SELF-HELP*

Third, it is attractive to the state if community involvement leads to some of the burden of programme delivery passing to local volunteers. And experience shows that local people are frequently more ready to give of their time, labour, expertise and money, not to mention spare seats in their cars, for example, or underused space in their buildings, if this is in a good, locally focused and locally determined cause. Not only is there a cost saving but, it is argued, more imaginative and innovative action may be forthcoming if the uniformity of a state-led approach is relaxed.

### *COMMUNITY DEVELOPMENT, EMPOWERMENT AND CAPACITY BUILDING*

Fourth, involving people as partners in decision-making and/or in programme delivery is also likely to increase and improve the 'human resource'. The argument is that involvement has an *educative or training function*, with individuals gaining new skills and awareness, as well as a *development function*, with local networks and institutions being spawned or strengthened and whole communities gaining the confidence and energy to launch new ventures at a later date, perhaps unrelated to the one in hand.

But what exactly is meant by the term 'community involvement'? Generally, it is used to denote some of the 'middle range' of activities in the spectrum or 'ladder' (Arnstein, 1969) of activities often evoked to denote a range of possible relationships between the state and the local community. Different writers tend to denote that spectrum a little differently from one another (see, for example, Warburton, 1998a and New Economics Foundation, 1998) but the main elements are as follows, with the strongest empowerment of local communities coming at the top:

1. citizen control
2. delegation
3. partnership
4. participation
5. consultation
6. education
7. information
8. manipulation

'Community involvement', in the present context, is construed to embrace levels 3, 4 and 5, namely *partnership*, *participation* and *consultation*. As Chapter 9 has already explored the concept and practice of 'partnership', we will confine ourselves here to:

- *consultation* – the process of seeking the views of local people; and
- *participation* – the process of involving local people in determining and/or delivering policy, programmes or projects (but without the state 'letting go' to the extent implied in the concept of 'partnership').

What has happened in recent years is that the point of interface between the local community and the local state – meaning, in general terms, the local authorities and the various local and regional quangos – has tended to 'move up' the spectrum set out above. This has arisen not just because the state has grown more and more keen to reap the four benefits listed earlier, but because the 'community' has grown in sophistication and has become less prepared to 'leave everything to the government'.

## SOME KEY ISSUES

A number of key issues surround the involvement of local people in the rural development process and in this discussion we are assuming that it is a local authority or some sort of local development partnership that is contemplating launching an 'involvement exercise' – though sometimes it is the local community itself taking the initiative, effectively appropriating to itself a measure of power.

The first key issue is *why* the involvement? Which of the four possible benefits listed earlier is the 'involver' hoping to achieve – better decision-making, more durable action, the promotion of self-help or the development of the community? Or is it just going through the motions with the key decisions already taken – an exercise belonging properly at levels 6, 7 and 8 of the spectrum? Certainly, a clear answer to this question will help to determine how the whole exercise is planned and carried out.

Going on from that, *who* should be involved? Is 'the constituency' just the people who live in the area in question – and how narrow or extensive is that? or should it include those who work or have businesses in the area as well as tourists and visitors? Does it comprise adults but not children, and all of the people or just those likely to make use of a particular facility such as a health centre, public transport service or youth programme? Is it a consultation of pre-existing groups or of the population at large? If the latter, will a sample suffice or a representative group of acknowledged spokespersons? This raises the whole issue of inclusiveness, of recognising that participants tend to be self-selecting and that the articulate will probably not reflect the views of excluded groups such as women at home, the homeless, unemployed people or ethnic minorities. There are ways of trying to involve such 'hard-to-reach' groups, but they take time and care.

*How* the local community is approached is also important. The purpose of the exercise needs to be explained and reactions sought at the outset. In seeking increased involvement there is a need to work through existing groups and institutions, to respect the roles and positions of local leaders, rather than to jump in unannounced.

There is a wide range of consultation and participation *tools* to choose from, and a clear temptation to get 'hooked' on a particular familiar one. 'Mix and match', or developing some sort of hybrid, is often possible and different techniques may be needed as a project progresses. For example, public meetings and the use of the media may be useful as a project is being launched, community appraisals and round-table workshops may be appropriate in the middle stages as popular opinion is being canvassed and ideas are being generated, and action planning may come into its own when a 'wish list' has to be translated into a viable plan.

Then there is the significance of the *rural context*. Many consultation techniques have been developed with an urban context in mind – a large housing estate or a run-down inner-city area perhaps. In rural areas, however, the wide scatter of the population provides its own challenge. Consulting the same number of people there as in an inner-city area may be impossible using formal meetings; indeed an insistence on consulting a similar number of people in a single exercise may involve basing it on a geographical area so extensive that it has no real meaning for those involved. And the use of three-dimensional physical models, as in 'planning for real' exercises, could well miss the point in a rural development context where the built environment may not really be at issue.

Often the careful deployment of a *community development worker* to initiate, facilitate and draw together the threads of a participation or consultation exercise is essential. But such a person has to tread carefully, respecting not just the sensitivities of the existing social and political structure in the area but also the need to let the local people themselves shape and steer the exercise to the maximum extent possible. This implies a sound training in the craft of 'facilitation'.

Then there is the need to remember that *involvement is a process*, not a one-off quick fix. Early disappointments need to be countered by a determination to build up a culture of participation in which people feel confident that they will be taken seriously in both fashioning and helping to deliver projects and programmes. In short, consultation is not just a matter of finding out what people want; it should be part of an ongoing process of mutual learning, partnership and the development of self-reliance. And 'participation' must involve a genuine readiness to allow people to do things differently and maybe make mistakes. All of this can be expected to take time, especially if a climate of demoralisation, elitism and fatalism provides the starting point.

## THE TOOLKIT

There are several compendia setting out a range of participation and consultation tools available and the pros and cons of each (for example Moseley and Cherrett, 1993; Environment Trust Associates and Local Government Management Board, 1994; DETR, 1997; Rural Forum and the Scottish Office, 1997; New Economics Foundation, 1998; Bur *et al.*, 1999). The tools vary greatly in their degree of sophistication and formality, in their capacity to embrace all social groups and in their ability to stimulate a real exchange of ideas and concerns rather than just an accumulation of individual opinions.

Many of the most popular methods today involve groups of people, possibly preselected to match a particular desired profile, coming together for a few hours, or possibly days, in order to 'brainstorm' or somehow seek to arrive at a common position or vision for the future of an area. Words like 'visualise', 'visioning' 'imagine', 'jury' and 'search' recur in the nomenclature of group consultation techniques. These groups may be based on clear sets of interests – such as those of farmers, trade unionists, conservationists or newcomers – or be intended to encapsulate a cross-section of the whole community. Other tools seek to embrace much larger numbers of people but may gain comprehensiveness at the expense of real debate and the testing of ideas.

All we can do here is set out in tabular form a small number of the possible methods indicating one or two strengths and weaknesses of each (Table 10.1).

**Table 10.1** ***Some approaches to consultation and participation in local development***

| What | In essence | Some strengths | Some limitations |
|---|---|---|---|
| Parish (or 'village' or 'community') appraisals | A questionnaire survey of, by and for the community, regarding its wants, needs, resources, problems etc. It leads to a report and usually local discussion and action points. ('Community appraisals' can also be taken to mean more comprehensive audits involving the use of much more than household surveys.) | • 100% of the community can participate<br>• anonymity<br>• good for identifying the nature and strength of local concerns | • opinions are not formed or honed in debate<br>• time-consuming and labour-intensive (often takes a year or more but survey software can help) |
| Modelling local sustainability | A gaming exercise in which small groups appraise possible local projects against environmental, economic and social criteria. The game can be board-based or computer-based. | brings together disparate interest groups or stakeholders to test each others' ideas and the merit of different options | • participation is by invitation only<br>• Requires high level of verbal skill |
| Local focus groups | Several variants – all tend to involve small groups considering key issue(s). Thus 'Future Search' involves stakeholders creating a shared vision via alternative scenarios. | contributes to local consensus building and an appreciation of trade-offs | • participation by just a select few<br>• requires high level of verbal skill |
| Parish maps | A large map of the locality is communally generated, locating whatever it is that local people cherish – an exercise producing a physical product | • inspires community awareness and spurs initiatives to conserve or enhance local features<br>• in practice, often involves women | may tend to favour physical, land-use and location-specific issues at the expense of the less tangible |

*(Continued)*

**Table 10.1** ***Continued***

| What | In essence | Some strengths | Some limitations |
|---|---|---|---|
| | as a stimulus to discussion. | | |
| Participatory rural appraisal | A composite term embracing various methods of learning from local communities, usually including semi-structured interviews with key informants, group discussions, gaming and ranking exercises to elicit preferences. | relatively quick way of identifying priority concerns and issues | requires careful preparation and good skills in community development |
| Planning for real | Local people use a three-dimensional model of the neighbourhood in a series of open meetings and reveal their needs by placing cards or flags in appropriate locations. A spur to discussion. | • use of a 3-D model is a good way of engaging people, e.g. re. road safety or building developments<br>• participants do not need high level of verbal skills | • strong physical, land-use bias<br>• best for a defined, built-up neighbourhood – less obviously useful for area-based rural development |
| Public meetings | Involves publicising and convening a meeting, open to all, with formal presentations by people in positions of responsibility, plus questions/ comments. | • open to all<br>• good for giving out basic information and raising initial awareness | • poor basis for real debate<br>• generally dominated by the articulate minority |
| Formal local democracy (parish councils and the like) | The periodic election of representatives by universal suffrage. The council then resolves a limited range of local issues but has authority to act and spend quite widely and also to represent local issues at a higher level. | • legitimacy that comes with legal status and formal recognition<br>• can be an effective focus for consultation and debate if the council is dynamic and open in style | • often has a poor image<br>• usually low turnout in elections (or no election at all) which raises doubts re. representative-ness<br>• can choose to work to a narrow agenda with little interest in local development |

## CASE STUDY 19: THE PARISH/VILLAGE/COMMUNITY APPRAISAL

The parish/village/community appraisal may be defined as 'a questionnaire survey of, by and for the local community, designed to identify local characteristics, problems, needs, threats, strengths and opportunities and thereby to create a sound foundation of awareness and understanding on which to base future community action'. The key words in that definition are 'of, by and for'. All three are essential elements, though the style and emphasis of parish appraisals does vary from place to place as should be the case with truly community-based endeavours. Indeed, the terms in common usage vary; there are parish, village, town and community appraisals, profiles and audits. But in the English context, most of the 2000 or so such studies that have been carried out in the past 25 years have been termed 'appraisals' and have related to civil parishes with populations of between 300 and 3000 (Moseley, 1997a).

For the various reasons underlying state support for community involvement that were reviewed earlier in this chapter, English local authorities and development agencies such as the LEADER II local action groups have shown an increasing readiness to promote and support parish appraisals in recent years. Sometimes they have agreed to defray a share of the costs incurred by local communities in carrying them out; sometimes they have undertaken to consider sympathetically any concerns or proposals that might emerge from them. But it is arguable whether the parish appraisal is strictly a 'tool of consultation' for the simple reason that normally the real impetus behind them comes from the local community itself and not from a 'superior' body anxious to consult.

Why this local enthusiasm for a task which is always demanding and generally takes a year or more of determined work to complete? One motive has been sheer curiosity, a wish by local people to research and publish a kind of local Doomsday Book or 'state-of-the-village' report. More common, however, has been a wish by some concerned group of people to explore which direction the community wants to go, sometimes in the face of a perceived threat such as a proposed large new housing development, sometimes with a general feeling that 'the place is going downhill', as evidenced perhaps by the closure of the village school or the steady decline of the number of young people. Sometimes there has been a wish to see if some factional view, perhaps that of a vocal preservationist lobby, really has majority support, or else to gather opinion in the hope of influencing the local planning authority as it prepares a new local plan.

What does an appraisal involve? The main stages are normally: working to establish broadly based support in the parish for the venture; forming a steering group to decide what, how, when and by whom; planning the survey and drawing up the questionnaire; collecting the information from the parish's households and/or individuals; analysing the information;

drafting an appraisal report to include recommendations and action points as well as the statistical evidence; distributing the report widely in the parish; local discussion to get a mandate to proceed; follow-up action; and monitoring and evaluation. Sometimes, as will be explained in Chapter 12, the appraisal can be just a preliminary step towards the preparation of a village action plan.

Thus the questionnaire survey is, or should be, just the centre-piece of a much longer and broader exercise. It *is* time-consuming and demanding, though in recent years help has been available in the form of computer software which offers local communities a long 'menu' of possible questions for their consideration, prints out the preferred questionnaire, handles data entry and undertakes data analysis and the presentation of the results in tabular and diagrammatic form (Countryside and Community Research Unit, 1998). None of this, however, is a substitute for local judgement at all stages of the appraisal exercise, and concern is sometimes expressed that the use of computer software risks turning a real community initiative into a technological fix.

There are two quite distinct views about the real value of parish appraisals. The first is that they are essentially a 'means to an end' i.e. a necessary precursor to some sort of tangible action on the ground such as the provision of bus shelters, play-groups, a community minibus or whatever. The second is that 'the process is itself the product' or, more accurately, that the real product is the enhanced awareness, confidence, resolve, skills and relationships generated by the appraisal exercise in the community as a whole and/or in many of its individual members. Happily, these two desirable outcomes are not mutually exclusive. Indeed, success in one can and often does breed success in the other, and the possibility of triggering an upward spiral of achievement in this way is precisely what community development is about.

But do appraisals genuinely lead to action? Research involving the author went some way towards clarifying 'what happened afterwards?' at least with regard to two counties in southern England, namely Oxfordshire and Gloucestershire (Moseley *et al.*, 1996). There, 44 appraisals carried out in the early 1990s were reviewed one to four years later and it was found that of the 422 separate recommendations or action points contained in the appraisal reports, roughly equal proportions had subsequently been wholly implemented, not implemented at all and – the intermediate category – partially implemented or else in some sense still ongoing. Looking at which particular recommendations and action points had met with most success, the most commonly implemented were those where the relevant power had rested largely in local, parish hands; the least successful were where agreement and expenditure were needed from superior bodies such as the county councils. Thus action points relating to better local information provision, the maintenance of footpaths and the establishment of 'good neighbour schemes' had very largely come to fruition, while those relating to traffic and road conditions, the provision of low-cost housing or the creation of new employment opportunities had proved much more intractable, though some clear successes were none the less evident even in those fields.

Several conclusions can be drawn from this and related research on parish appraisals:

- Parish appraisals have been successfully undertaken by a wide rage of local communities, of different size and social composition and with different characteristics and problems; the 'triggers' to undertake appraisals have varied greatly.
- The typically high response rates achieved in the surveys are a testament to the level of interest people have in their local community when asked to express an opinion, and they provide a reasonable mandate for subsequent action.
- Virtually all appraisals have generated some beneficial outcomes relating either to action on the ground and/or to the wider domain of 'community development'.
- As far as 'action on the ground' is concerned, success tends to depend on the dynamism and determination of a few individuals in the community; the level of appreciation locally that follow-up needs to be carefully planned and vigorously pursued; the degree of support and enthusiasm coming from the elected parish council; and the support of outside agencies such as the local authorities and the county-based Rural Community Councils.

In short, parish appraisals are an important social innovation which has helped to devolve to local people a significant share of the task of caring for their local community and environment as well as giving statutory agencies and voluntary bodies a firmer factual basis upon which to plan their services.

But three cautionary notes should be sounded: first, they are much better at crystallising a community's needs and wants than the community resources that might be harnessed to help meet them; second, they may imply an element of 'first up, best dressed', with those communities most adept at getting themselves organised winning a disproportionate share of any resources on offer; and third, response rates of 60 to 80 per cent are impressive in any household survey but the silent 20 to 40 per cent should always be borne in mind as they may well comprise most of the truly disadvantaged people in the community.

---

## CASE STUDY 20: WALLONIA'S COMMUNE PROGRAMMES FOR RURAL DEVELOPMENT

---

As a continental example of good practice in involving local people in rural development we may cite the work of the Fondation Rurale de Wallonie, which serves the 130 or so essentially rural communes of Belgium's French-speaking Walloon region. The Wallonia regional government has responsibilities that include economic development, strategic land-use planning and environmental management, as well as the supervision and part-funding of the commune councils. As for the latter, they too enjoy considerable responsibility including the provision of a local police service, road maintenance and ensuring the provision of water, gas and electricity – but with the more rural communes typically having only about 7000 inhabitants they often have to act in concert with their neighbours or other agencies.

For more than 20 years the Wallonia regional government has sought to address such rural issues as agricultural restructuring, environmental pressures, the need for a broader employment base and the difficulties of servicing a scattered population. Building on its early experience, it legislated in 1991 to give force to a number of basic principles of rural development and, more specifically, to a 12-stage programme that communes would be obliged to follow if they were subsequently to be eligible for the substantial financial assistance potentially available for rural development projects.

The guiding principles, which have been firmly adhered to, are:-

- development should be planned and managed at the commune level and by means of a Commune Programme for Rural Development;
- these programmes must be integrated, covering most or all aspects of life – the local economy and employment, housing, transport, environmental enhancement, service provision, community development etc.;
- the programmes must be designed so as to make fuller use of local physical and human resources; and
- the production and execution of the programme must involve local people fully; consultation and citizen participation must be central to the whole process.

It is the last of these – the requirement to involve local people fully, as expressed in a particular commune – that we are most concerned with here. But first it is necessary to explain the key role of the Fondation Rurale de Wallonie (FRW) which was set up in 1975 to foster rural development in the region. Though a not-for-profit organisation with a social purpose, it is run on quasi-commercial lines with the regional government and the communes being, effectively, its clients. From 1979 onwards it piloted and refined a rural development process in partnership with over 20 communes, a process which was later to be enshrined in law, as explained earlier.

Much of FRW's work on the ground is carried out by about 40 *agents de développement* whose task is to help commune councils and their residents to prepare and implement the rural development programmes. By 1995, 46 of Wallonia's rural communes had contracted with FRW to work together in this way, these communes containing some 500 individual villages and over 300,000 residents. So there is now a tried-and-tested procedure for rural development in Wallonia set out in an unambiguous legal framework. Information, consultation and participation are formal requirements and minimal procedures are laid down in the 1991 law.

In 1995/96, the author was able to study these procedures as followed in the Walloon commune of Brunehaut, as part of a transnational project relating to community involvement in rural France, Britain and Belgium (Fondation Rurale de Wallonie, 1996). The commune of Brunehaut comprises nine former communes near the French border, amalgamated in 1977 and covering 46 sq. km, with a population of some 7500 people. The commune council, elected every six years, has 19 members of whom a group of five, chaired by the mayor, comprises its executive. The council employs some 60 staff and has an annual expenditure in excess of £1.5 million.

In 1990, when the commune council resolved to launch a rural development programme, the commune had a 15 per cent unemployment rate, employment in both manufacturing and agriculture in sharp decline, earned incomes well below the regional average, a falling school-age population and the imminence of reform to the Common Agricultural Policy causing some anxiety among the commune's farmers.

The commune's rural development programme went through four phases.

### The information phase

In the autumn of 1991 a two-hour, peak-time regional television presentation and debate about the proposed Brunehaut rural development programme was seen by half of the commune's residents, many of whom phoned in with questions and observations. After that, four local meetings, scattered throughout the commune, were convened to discuss the initiative, and an exhibition was staged in the offices of the commune council. Together, these initial ventures persuaded about 70 people to volunteer to participate in the working groups which would carry things forward.

### The consultation phase

Seven working groups were established and they met a total of 18 times between November 1991 and May 1992, averaging ten participants per meeting. The groups focused on topics suggested at the public meetings, namely the economy, tourism, housing, agriculture, social and community life, the environment and road safety. Questionnaires were distributed to a sample of residents to ascertain the perceived strengths and weaknesses of the commune for discussion in the working groups.

### The local commission phase

In April 1992 a Local Commission for Rural Development was established in the commune as required by the 1991 legislation. Chaired by the mayor, and with its members drawn widely, both geographically and socio-economically, from lay people within the commune, the Commission's role was to be a 'permanent consultative body at the disposal of the commune' charged with drafting the Commune Programme for Rural Development and with assisting its subsequent implementation and monitoring. During 1993, it duly worked to produce the Draft Plan, with technical assistance from the FRW and drawing on the opinions and proposals that had come from the various working groups. The Draft Plan set out about 60 specific projects together with an indication of their relative priority, cost and relationship to relevant objectives, and went forward for the approval of the commune council.

**The implementation phase**

By November 1993, the commune council and the Wallonia regional council had both approved the Plan, and in January 1994 the region announced that it would fund 80 per cent of the cost of five projects given high priority in it. These related to the creation of a 'shop window' for local produce, the conversion of a redundant building into a village hall-cum-environmental interpretation centre, the redesign of a congested square in one of the villages, traffic-calming measures outside two village schools and the conversion of another redundant building into a village hall incorporating a unit of social housing. By early 1995, all five of these projects had come to fruition together with four smaller schemes set out in the Plan and wholly funded by the commune. The working groups were then reconvened by the Local Commission to bring forward further proposals for its consideration and possible inclusion in a rolled-forward Plan.

Many useful conclusions about rural community development can be drawn from this experience. It was genuinely multi-sectoral, dealing with issues defined as important by local people. The consultation exercises were genuine, and well-planned and executed. A precise legal framework for that consultation removed the temptation to 'cut corners'. The continuing support of an experienced rural development agency deploying a specific officer was crucial to the project's success. Local effort and enthusiasm was sustained in large part by the knowledge that the region's financial support for selected projects was assured so long as the job was done properly. And finally, the involvement and authority of the commune council, and the Local Commission that it established and owned, firmly anchored the process in the locality and gave it legitimacy.

From its launch on regional television to the completion of the first tranche of projects the process took about three-and-a-half-years. Over and above the tangible successes on the ground, less obvious but equally valuable benefits had been secured. These included an enhanced sense of belonging and local pride, a reduction of suspicion and rivalry between the constituent villages of the commune and a flourishing of active citizenship.

### *SEE ALSO...*

Several other case studies relate directly to community involvement in local development, especially numbers 2, 5, 12, 24 and 28.

### *SELECTED FURTHER READING*

DETR (1997), Derounian (1998), Lowe *et al.* (1998), New Economics Foundation (1998) and Chapter 12 of the Rural White Paper (Departmant of Environment, Transport and the Regions and MAFF, 2000).

# Part II

## Practice

# 11

# Diagnosis: Researching the Baseline

'Get your facts first and then you can distort them as much as you like.'

Mark Twain

Having examined the overarching themes of local rural development we now continue with four chapters devoted to the practicalities of preparing and executing strategies to achieve such development, considering in turn the stages of diagnosis, planning, implementation and evaluation.

## DEFINITIONS AND IMPORTANCE

Logically, *diagnosis* is the first step in this process. It is the task of establishing the 'baseline conditions' of an area prior to creating a development programme for it – this baseline to include reference to resources, opportunities, problems, needs, constraints and the relationships between them. Another term for it is 'baseline study'.

The importance of this exercise has been underlined by a number of rural development agencies. For example, England's Rural Development Commission, launching the preparation of a new round of locally focused Rural Development Programmes (RDPs), made clear that

> an essential part of the RDP process will be the production of a strategy which identifies needs, opportunities and resources in the Rural Development Area in question...it should be based on an assessment of social and economic trends and conditions and the needs and opportunities to which they give rise. There should be an assessment of the resources likely to be available...the areas of greatest need, any particularly disadvantaged groups and the most serious problems in the area. (Rural Development Commission, 1994: 12)

It went on to give advice to its various local partners on the scope, method and source material of the diagnosis, and stressed that 'getting this initial analysis right is vital to the success of the RDP process'.

In the case of the European Union's LEADER II programme, technical assistance was given to new and inexperienced local action groups so that they could undertake appropriate preparatory work, this to include analysing the needs, wishes and motivation of local people, the formation of local partnerships and the preparation of a strategy and business plan. To this end, a technical paper suggesting how best to plan and execute a 'diagnosis of the area', and distilling good practice from recent experience in France, Spain, Portugal and Britain, was produced by an international team, including the present author (Chanard *et al.*, 1994).

All of this reflects a measure of disappointment with the quality of earlier baseline studies (Moseley, 1996b). For example, an analysis of the LEADER I programme, which was introduced in great haste in late 1991 and early 1992, reveals many cases where the initial survey of needs and resources was skimped or was, effectively, non-existent. Informally, local LEADER groups asserted that 'they knew what the problems were' and that in their view further analysis would simply hold things up.

This rather complacent view, however, has been challenged frequently. For example, an official evaluation of the 16 Irish LEADER I programmes concluded that 'most of the Irish groups are weak on strategic planning. More emphasis should be placed on comprehensive resource audits, detailed SWOT analyses (strengths, weaknesses, opportunities and threats), prioritisation and the identification of people with leadership and management skills' (Kearney *et al.*, 1994: 133).

More recently, England's Countryside Agency has urged local partnerships keen to turn round the steady decline of many market towns, to develop action plans based on careful 'health checks' of the towns involved, and below we summarise the essence of the recommended approach (Countryside Agency, 2001c).

## SOME KEY ISSUES

Should the approach be *technocratic* or *participatory*? A first issue concerns whether the diagnosis should be mainly the reserve of an expert team, working largely from statistical and other documentary evidence pertaining to the area, plus some interviews with a small number of 'key players', or whether a substantial effort should be made to get local people to articulate their opinions and, indeed, to serve as 'witnesses' of the local scene.

Frequently, local partnerships or development agencies need to be persuaded that a more participatory approach is beneficial, but obviously this is not a straightforward matter of either/or. Each approach has its contribution to make, and the relative weight attached to each may reflect the availability of appropriate statistics and secondary material, including any previous collations of popular evidence and opinion such as earlier 'community appraisals', as well as the readiness of local people to respond to invitations to get involved. This point is returned to in our consideration of the 'toolkit', but it should be stressed that, when set alongside the 'technocratic' approach, the deliberate involvement of local people as sources of intelligence promises to yield more than just useful information. It also promises greater legitimacy for the whole development programme when it finally emerges and, quite probably, greater or more widely based enthusiasm for its execution. Indeed, a diagnosis with a strong participatory component can quite properly be seen, and indeed justified, as a part of the development process itself (see the previous chapter on community involvement).

Another key issue is *scope* – what topics does the diagnosis need to cover? Obviously, this depends on what information and understanding is needed before a medium- or long-term strategy can be drafted, setting out the development agency's goals and objectives and the main elements of the programme that will be mounted in order to attain them.

Thus the diagnosis has to be based on 'the need to know' principle; there simply is not the time to be comprehensive or, indeed, to engage in a lot of original research. Happily, 'original research' – for example into soil quality or the local industrial structure – is rarely needed in western Europe; other agencies exist to research and make available such information. A central task of the diagnosis is to sift through the available information for evidence of underused resources and potent constraints which would hold back a development programme. For most locally focused rural development programmes three or four months may be about the right period of time to spend on the diagnosis phase; much less and superficiality is risked; much more and there could be a loss of momentum with an erosion of interest in the whole venture.

Another key point concerning the scope of the diagnosis is the importance of considering not just phenomena that can be clearly seen on the ground or easily captured in statistics, but the invisible and often intangible as well, such as the context provided by public policies, the local culture and popular concerns and aspirations.

In reality, the scope of useful diagnoses for rural development reduces to two main themes. First is the *good news* comprising the area's resources, opportunities and strengths which may best be considered under such headings as:

- its *natural resources*, to include its location *vis-à-vis* potential markets, its natural environment, climate, soil etc.;
- its *capital endowment* including its infrastructure, built environment, cultural landscape etc.;
- the *human resource*, meaning people as individuals, their skills, leadership capacity etc.;
- the *social and institutional environment*, including the range and vitality of small businesses, the existence of voluntary organisations and the quality

of any existing partnerships bringing together agencies with an interest in local development;

- the *policy context*, meaning the existing position of the relevant agencies with regard to the area's development potential. Such positions could comprise an opportunity, a constraint or a mixture of the two; and
- *possible future scenarios*, which could again provide opportunities to be exploited. A new motorway being built in the vicinity or the growing popularity of short, second holidays out of season provide examples.

Second is the *bad news*, meaning problems, weaknesses, constraints and threats which would need either to be carefully side-stepped or else tackled directly so as to reduce their force. The categories listed above under 'the good news' are equally relevant here though the task now is to check for weaknesses – to see, as it were, in what respects 'the bottle is half empty rather than half full'. Included here also, though arguably not 'bad news' as such, are the expressed problems and needs of the local population and business world. It is clearly important that these be established early on with a decision taken, again, on whether the development strategy can and should seek to address them or else 'put them to one side'.

The scope of the Countryside Agency's 'market town health check', referred to earlier, is a broad one (Countryside Agency, 2001b). In essence, the local partnerships involved are urged to complete 18 'worksheets' containing some 180 questions which span the town's environment, economy, community and accessibility. Typical questions are: 'Is the local economy dependent on a small number of large employers?' and 'How seasonal is visitor activity?' Advice is given on the variety of sources that might be tapped for useful information and on how the completed worksheets might then be analysed as a springboard for local debate and subsequent plan-making.

Obviously, there are various *dimensions* to such diagnoses. For area-wide development strategies there is obviously a need to assemble and appraise some information geographically and to map the various resources, constraints, needs and problems. But there is equally a need to address the *non*-spatial dimension of the key issues and phenomena – for example, their impact upon different social groups or economic sectors present in the area. Again, a multi-dimensional perspective is likely to be best. In any event, the time dimension is important. Diagnostic exercises should seek to establish trends and the dynamics of change rather than simply construct a snapshot of the position 'today', with 'today' often meaning the recent past because of the time-lag in information becoming available.

## THE TOOLKIT

We now move on to consider some of the tools or approaches that might be employed in carrying out a diagnosis of an area prior to formulating and executing a development programme. Obviously, the best diagnoses will probably comprise a mixture of the different tools which are not mutually exclusive.

### *STATISTICAL MAPPING*

Generally, this approach tends to comprise the assembly, analysis and mapping of small area statistics from the most recent census and other sources. For example, one French method (Commissariat Général du Plan, 1991) involved mapping at the commune level, across wide rural areas, the incidence of *fragility and vulnerability* as an indicator of the local *need* for development, and of *local dynamism* as an indicator of the local *potential* for sustaining development.

In all, this approach involved about 60 individual measures being collated and mapped. As far as 'fragility and vulnerability' were concerned, one example of these measures was social isolation, as revealed by the percentage of one-person households and the unemployment rate; and concerning 'local dynamism', the number of building permits issued in the recent past per 1000 inhabitants, and the proportion of households sufficiently prosperous to be liable for local taxation, were relevant indicators. In addition to readily quantifiable measures of this kind, others based on local judgement were used – for example, in the case of 'fragility' any perceived mismatch between the training which was on offer locally and the needs of local employers; or, in the case of local dynamism, how far local service providers were availing themselves of the new information technology.

A limitation of this approach, when used in isolation, is that it leaves unresolved a central question, namely whether development initiatives should reflect the geography of *need* or that of development *potential*. Or, to put it more succinctly, should the subsequent development programme put 'the worst first' and channel resources to 'black spots,' or should it 'back winners'? Resolving that dilemma generally involves 'getting under the skin' of an area rather more than is achieved by the mapping of indicators or indeed by statistical analysis more generally.

### *THE WRITTEN WORD*

A second approach involves the systematic scrutiny of a whole range of *documentary evidence* relating to the locality – such evidence being generally plentiful in most of western Europe. The challenge is partly to track down what exists and partly to know how to extract what is helpful. Regarding the latter, one approach is to begin with a series of worksheets with headings such as 'resources, constraints, popular concerns, opportunities, threats' and to condense the 'messages' of the texts accordingly.

Going back over, say, the previous three years, relevant material will probably include:

- minutes of the meetings of local elected councils and of major voluntary or community bodies;
- any planning studies undertaken for, say, the land-use planning authority or the health authority;
- local newspapers, concentrating perhaps on the editorials and letters pages; and
- any student dissertations or theses.

Obviously, it is a method fraught with dangers, since the representativeness of the views expressed and their grounding in 'fact' – itself a slippery concept – must always be borne in mind. But used alongside other methods, and particularly used as a way of drawing up a detailed agenda of local issues for debate, it is a very useful tool and one which is easily overlooked by those who are minded always to undertake 'yet another survey'.

### THE RESOURCE AUDIT

Another approach focuses explicitly on one key element of the situation – the area's resource endowment. One way of attempting this was again set out in the French paper referred to above (Commissariat Général du Plan, 1991). It involves quite simply posing six questions in relation to each of six sets of resources; in effect, the research task is to fill in each cell of a six-by-six matrix. Clearly, the focus here is upon the economic development of the area – a rather different matrix and associated questions would be needed if the emphasis was on social development. And it should be said that little guidance is given on precisely how to 'fill in the matrix' though it would seem that a set of structured interviews with appropriate 'resource managers' to distil their views would be essential.

The six sets of resources are:

- *natural environment* – the land, forest, water, hills, fauna and flora;
- *cultural heritage* – the landscape, historic monuments, local architecture, local products and know-how, local culture, local brands and trade marks;
- *infrastructure* – the area's communication nodes, its proximity to urban centres, serviced industrial land, housing available for rent;
- *people* – their capacity for innovation, leadership qualities, the number of natives of the region in positions of responsibility, and people with specific qualifications;
- *organisations* – the machinery for inter-communal co-operation, strong voluntary bodies, strong collective structures such as co-operatives, subsidiaries of major companies and the presence of regional labels; and

- *financial resources* – the capacity of business for investment, domestic saving, the fiscal resources of the communes and the local tax base.

To be applied to each of these are the following six questions:

- Is there a market capable of exploiting this resource?
- What actual or potential competition from elsewhere exists?
- Is existing local machinery capable of exploiting this resource, and under what conditions?
- How far is this resource actually being exploited?
- How urgent is its exploitation (or its conservation)?
- Is further evaluative work needed?

### THE CULLING OF EXPERT OPINION

Another 'good practice guide' on baseline studies emanating from the French government (Ministère de l'Equipement des Transports et du Logement, 1997) advocates the creation by the local development agency of a widely based 'forum' of around 50 to 150 local leaders, with a mixture of qualitative personal interviews and group discussions being employed to elicit from them their list of key local issues and the essential factual information pertaining to them. Increasing these leaders' ownership of the subsequent development strategy was said to have been a valuable spin-off from this particular fact-finding approach when it was piloted in France.

More generally, we may mention SWOT analyses, 'SWOT' standing for Strengths Weaknesses Opportunities and Threats. This technique involves getting well informed local people and others from outside the area who are well-informed on some aspects at least of the local situation or of the forces acting upon it to come together and reflect, not alone but collectively, on those four key dimensions of the local situation. How this might be undertaken in practical terms is suggested in the first of the case studies below, so we need not elaborate here.

### LOCAL TESTIMONY

Finally, we must refer to the systematic assembly of evidence and the culling of opinion from a much wider sample of the local population. Several such techniques were covered in the 'Community Involvement' chapter; particularly valuable in the present context are those which incorporate a comprehensive or sample survey of the residents of the area under scrutiny. But this is not without problems given the need to proceed quickly, as explained above. There are many examples of 'village appraisal exercises', for example, that have been launched with great enthusiasm by local activists as a way of establishing local 'testimony' to be fed into a development planning exercise, only for the whole process to take so long that the opportunity to influence events is missed.

More realistic, therefore, may be the scrutiny by the research team of the crop of local appraisals or other social surveys recently carried out in the area, with a view to distilling useful evidence. If such reports do not exist, and if a three- or four-month time limit is effectively in force, then the best approach may instead be to convene 'focus groups' or something similar involving representative samples of 'ordinary local people' – again, the chapter on community involvement refers to this.

## CASE STUDY 21: A BASELINE STUDY FOR THE FOREST OF DEAN RURAL DEVELOPMENT PROGRAMME

In 1994 the author was commissioned to carry out a baseline study prior to the preparation of a comprehensive development strategy for the Forest of Dean Rural Development Area which lies on the England–Wales border in the county of Gloucestershire (Moseley, 1996b).

Briefly, the Forest of Dean contains about 75,000 people spread over some 500 sq. km. It is a former coal mining area, more of it forested than farmed. In the mid-1990s it had above-average unemployment, a vulnerable manufacturing base, a weak service sector, an ageing population, four small towns, some straggling ex-mining villages, a strong sense of cultural identity and concern for 'a youth problem', with petty crime, drug abuse and a general feeling of disaffection being frequently cited in this respect.

Another consultant was retained to explore the area's resource base – meaning its financial resources, its physical infrastructure, the availability of office and industrial floorspace and, more generally, the area's assets when compared with its competitors. The present author was asked to concentrate on 'local need', with a brief to examine the nature and severity of social and economic need in the area and the appropriateness of targeting resources at selected problems, sub-areas and social groups. Thus, from the start, there was an acceptance that, given scarce resources, explicit priorities would need, in due course, to be established on three 'dimensions', namely issues, geography and social structure. In effect, the author had two months to crystallise advice on those priorities.

Six exercises were undertaken, with that aim in mind:

First was *a review of recent literature* relating to the area, notably some recent papers on possible economic strategies, the local authority's draft corporate plan, a recent survey of business opinion, the county council's structure plan and rural strategy, the district local plan, academic studies of housing need and of personal accessibility in the area, a government report on agricultural change, and papers by the county's Training and Enterprise Council on various labour market issues.

Second, with colleague Peter Gaskell, *an analysis of 1991 census data* was undertaken, focusing on 27 indices of social vulnerability and deprivation at parish level, in an attempt to elucidate both the 'geography of disadvantage' and the existence and size of disadvantaged social groups. This analysis revealed, for example, that the four small towns in the area had a disproportionate share of its unemployed people and lone-pensioner households, and also that the area as a whole had 700 households with dependent children, but no car, and 2100 elderly people with both a limiting long-term illness and no car.

Third came *interviews*, in person or by telephone, with 26 leading local figures in the public or voluntary sectors, each having responsibility for some area of policy or service delivery in the area.

Fourth was a synthesis of seven *village or town appraisals* which had been undertaken by local people in the preceding five years, these household surveys having, between them, gathered evidence from 16 per cent of the area's population. Each appraisal pointed up local hopes, fears, likes and dislikes.

Fifth, an original *household survey* was undertaken, involving in-depth discussions with about 150 households drawn at random from a council housing estate in a small, rather run-down town, an ex-mining village and a more prosperous commuter/retirement village.

Finally, six consultation meetings were convened, involving discussions based on draft papers about the area's situation, namely:

- a SWOT analysis, and
- five public meetings in different villages to discuss draft proposals.

The **SWOT analysis** involved bringing together for half a day about 30 people from the private, public, voluntary and community sectors in the area, casting the net wide to try to produce a good socio-economic, age and gender mix. A proviso of the selection of invitees, drawing upon local advice, was that each should be well connected with a range of other local people, for example as a trade unionist, elected councillor, local employer, journalist or youth leader.

After a short 'warm-up' session, explaining the general purpose of the development exercise in prospect, the participants were divided into four groups each of seven or eight persons, and despatched to separate rooms with a specific brainstorming task to be undertaken with the help of a flipchart.

These four tasks, one per group, involved: '*Thinking about the Forest of Dean and looking forward over the next few years, what do you see as its main strengths (or weaknesses/opportunities/threats) that could help determine its future*?' Then, after about 90 minutes, the four groups were brought back together. Their four lists were displayed and explained with reference to the discussion that had taken place earlier. Further collective discussion ensued; the four lists were extended, modified, rewritten and qualified with caveats, and finally a broad consensus emerged with regard to each. The composite list, with 22 elements, was typed up and went on to form an important component of the diagnosis of the area and a plank upon which policies and initiatives were developed.

A few examples from the list derived by the SWOT analysis may be mentioned. Among the *strengths* and *opportunities* were:

- a population density, high by rural standards, which opened up the possibility of considering a large part of the area as a 'single, dispersed city' of 60,000 inhabitants – with important implications from the labour market, retail market and transport viewpoints;
- the area's location within a 'motorway box' which puts the Bristol, South Wales and West Midlands conurbations all within 90 minutes' driving time; and
- an undervalued industrial heritage, including mines, quarries, redundant railway lines, tram roads, canals, watercourses and spoil heaps.

Among the *threats* or *weaknesses* identified were:

- gaps in infrastructure relating particularly to the internal road network, the lack of a main-line railway station and the lack of up-market hotel accommodation;
- a certain 'cultural resistance' to training and retraining among certain social groups; and an
- over-reliance upon a few sectors of the economy, notably metal manufacturing, and upon a few large firms.

Elements from the baseline report which was distilled from these six rapidly undertaken research exercises fed directly into the strategy document and, from that, into practical projects funded as part of the subsequent annual operational plans.

## CASE STUDY 22: THE DIAGNOSIS OF TOURISM POTENTIAL: A EUROPEAN MODEL OF GOOD PRACTICE

Our second case study focuses not on an all-embracing diagnosis of the development potential of an area but on establishing the local potential of one particular sector, namely tourism. In so many of Europe's rural areas tourism is seen as the main source of economic salvation but the decision to put tourism centre-stage in a local development programme is often based more on hunch and a sense of 'What else can we do?' than upon firm analysis.

With that in mind, the LEADER Observatory brought together in 1996 a multinational group of rural tourism operators and consultants, together with a number of local development managers with experience of the challenge of promoting tourism initiatives, and what follows is a distillation of the advice that emerged from that seminar and from the 'guide to good practice' that was subsequently published (LEADER Observatory, 1997d).

Five stages of a 'diagnosis of local tourism potential' are proposed in that guide, drawing on a range of data sources including local tourist guides, brochures and reports; site visits to the major attractions; surveys of at least a sample of the main suppliers of accommodation and facilities; a survey of each commune's resource base and of the issues perceived at that level; and a perusal of the literature of neighbouring competing areas as well as some located further afield.

**An analysis of supply**

Obviously this should include defining and mapping key natural resources such as lakes and distinctive landscapes, and cultural resources including local festivals, craft producers and historic buildings, as well as the availability of leisure facilities, accommodation, conference centres, catering capacity etc. But the notion of 'the local tourism resource' has to be interpreted more widely to embrace the existence and role of local tourist organisations, current marketing and training activities related to tourism and, more generally, the attitude of local people to tourism development – would they be positive and supportive or essentially opposed?

**An analysis of demand**

Here the need is, first, to learn about current customers visiting the area, whether as tourists staying at least one night and therefore clearly spending some money in the area, or as day-trippers. Where do they come from? What types of people are they? Where do they go and what do they spend their money on? What were their expectations on arrival? What do they like and dislike? Some such information can generally be assembled by desk research, but at least sample surveys of tourists will be needed, whether at their place of accommodation or at particular tourist attractions. However, it is important not to overlook other possible sources of useful information such as car registration plates and the accommodation registration forms of visitors. Also valuable are wide-ranging round-table discussion groups bringing together representative small samples of tourists to explore their expectations and experiences.

**An analysis of the competition**

Once the area's own tourism products are reasonably clear, it is time to gather information on existing and potential competing areas. Four questions are of central importance, namely: What are the main competing areas? What tourism products do they market? What are their weak points and strong points? How can information on these competitors be used effectively? This data-gathering exercise should focus on the immediately neighbouring areas – important as far as day-trippers are concerned – and other areas, further afield, offering similar products. Such analysis of the competition can lead to three alternative strategies:

- *imitating the competition* – by reproducing some successful feature;
- *being inspired by the competition* – to develop a product or service or idea in a different form; and
- *standing out from the competition* – by concluding that a market niche exists and seeking to exploit it.

**An analysis of trends**

The next step is for the 'diagnosis' to explore relevant external trends, particularly those affecting consumer behaviour. It is important to make maximum use of existing studies, though some local analyses, perhaps involving focus groups of local holiday-makers and tourism operators, can be useful in helping to project the local significance and relevance of 'global trends' distilled from the literature. Some trends which could affect the viability and nature of 'rural tourism' in the twenty-first century include: the ageing of Europe's population; a decline in mass tourism in favour of more differentiated products; a growing interest in environment and in health; and a growth of shorter, more numerous and often last-minute holidays. The task is to draw up a list of pertinent trends of this sort and relate them to the local situation.

**Putting it all together**

The final stage involves relating the previous analyses of supply, demand, competition and trends to one another with the aim of discovering the area's strengths, weaknesses, opportunities and risks as far as the development of tourism is concerned. Really there is no rigorous and clear-cut way to do this, though the LEADER 'good practice guide' essentially reaffirms the example described earlier in relation to the Forest of Dean – first, sift through the output of the earlier exercises to draw up a tentative listing of strengths, weaknesses, opportunities and threats/risks, and then use that list as a stimulus for discussion in 'think-tank' or 'open forum' situations, bringing together experts and tourism professionals as well as samples of the local population and of tourists visiting the area.

The intention is that from this exercise will emerge the area's *successful strategic position*, and also suggestions of the action needed to exploit it.

A specific example described in the same publication (LEADER Observatory, 1997c) may be useful. It concerns a local LEADER group which was keen to promote appropriate tourism in the Rhon Massif and followed the suggested diagnostic approach with the help of a firm of tourism consultants. The Rhon Massif is a well-defined mountainous region of about 2000 sq. km spanning three Lander in the centre of Germany and enjoying UNESCO 'Biosphere Reserve' status. The diagnosis involved a survey of 85 hotels, visits to 44 communes, interviews on site with about 600 tourists, various 'desk analyses', for example of literature relating to competing areas, and a number 'open forum' discussion groups from which many valuable conclusions emerged. For example, when the existing advertising material was analysed, it emerged that many tourist operators and promotional organisations simply failed to mention the area's 'unique selling point', namely its 'biosphere reserve' status. And, encouragingly, 75 per cent of visitors said they were prepared to pay some sort of modest 'environmental tax' as a way of funding conservation work, possibly as a levy added to the cost of overnight accommodation or to car-parking charges.

From such survey findings, and from work undertaken in discussion groups on strengths, weaknesses, opportunities and risks, a plan for the development of tourism eventually emerged with a broad measure of support. It defined a number of target groups, for example relatively elderly but 'young-at-heart' people and visitors preferring short but frequent holidays, and, going on from that, over 200 'action points' including developing the Rhon Massif's 'biosphere reserve' label and encouraging local restaurateurs to include more regional dishes on their menus. The whole point was to identify measures that would be mutually supportive and coherent, and that objective could not have been achieved without having first taken a rigorous, holistic look at the potential of the area as a whole.

---

All the evidence suggests that baseline studies of the kind described in this chapter repay the effort involved and offset any frustration arising from not being allowed to 'just get on with it'. Not only are evidence and ideas assembled which will provide a sound basis for action, but many of the people involved in the diagnosis are likely to be affected by the experience in a way that renders them more valuable in the process of local development. The following chapter explains how both of these products of the diagnosis (better understanding and capacity building) may be exploited in the subsequent planning process.

### *SEE ALSO ...*

Several other case studies touch on diagnosis, notably number 7 which deals with the adding of value to identified resources; numbers 19 and 20 which focus on community involvement; and numbers 23 and 24 which suggest how such diagnoses underpin strategic planning.

### *SELECTED FURTHER READING*

Moseley (1996b) and Chanard *et al*., (1994) provide overviews and some practical suggestions. The Countryside Agency (2001b) guidelines on assessing the development potential of England's small towns has wider applicability.

# 12

# Strategic Planning: Orchestrating Action

> 'If you would hit the mark, you must aim a little above it; every arrow that flies feels the attraction of the earth.'
>
> Henry Wadsworth Longfellow.

## DEFINITIONS AND IMPORTANCE

Here we are concerned with local 'strategic planning', 'strategic plans' and their value in local rural development. The previous chapter considered diagnosis, which provides a necessary point of departure for strategic planning, and subsequent chapters will address the implementation of plans and the evaluation of their progress and achievements.

Local strategic planning is 'a structured process whereby the actors in a local area define a long-term vision for the area based on identified needs and priorities, and set out clear objectives and a range of measures to work towards that vision over a defined period of time' – a definition based on Flanagan *et al.* (1995). A strategic *plan* is the product of that process, setting a framework for more detailed operational plans and programmes. Note that neither sort of plan need have a land-use component, though town and country planners normally conceive plans in that sense. Note also that it is not the geographical extent of the area in question that determines whether or not a plan is 'strategic' but the degree to which it stands above the detail. That said, this chapter is written with the challenge of planning a development programme across a reasonably extensive area – such as a LEADER area – very much in mind.

The importance of strategic planning is that it seeks to distil answers to three key questions: What do we want to do? How are we going to do it? How will we know if we are doing it right? Any strategic planning exercise can be judged by the rigour and care with which it seeks answers to those questions and the clarity with which the subsequent plan expresses them.

Not surprisingly, an enormous amount has been written on strategic planning since it is relevant to almost any human endeavour – government, business, public service delivery, military action, or whatever. But recent years have seen the production of a growing literature specifically designed to guide those engaging in local rural development (e.g. LEADER Observatory 1995b; Area Development Management, 1995), this literature seeking to apply the general principles of strategic planning in that context.

This reflects, perhaps, certain particular characteristics of local development when compared with those other endeavours, notably:

- its multidimensional scope – social, economic, cultural, environmental and political – and the need to integrate those concerns and dimensions;
- the multiplicity of 'actors' or 'stakeholders' involved, 'stakeholders' being persons or organisations with a legitimate interest in the matters considered;
- the need to treat the planning process as an exercise in involvement and capacity building – in other words to see the process as itself one of the engines of development; and
- the requirement that genuine *development* ensues and not just the better delivery of certain goods or services.

And so below we consider strategic planning as practised in pursuit of local rural development. Generally, the assumption will be that a development agency or local partnership at the 'area-wide level' is undertaking this task but, increasingly, strategic planning for local development is being attempted by very local, often *ad hoc*, groups working at the level of the individual village, parish or small town (Murray and Greer, 2001; Owen, 2002). Indeed, the second case study in this chapter relates specifically to that context. The general principles of strategic planning and the available tools are broadly valid at either scale.

## SOME KEY ISSUES

In the context of local rural development, strategic planning is designed to produce three outputs. First, and most obviously, a *document*, a 'plan' to guide subsequent action, as described below. Second, some *value added* to the actors and institutions involved; this is likely to come about from the mutual learning and debate which is intrinsic to the planning and bargaining exercise. Third, a shared *commitment* to the vision and the measures agreed upon and set out in the plan. This commitment, which can be informal or formalised in various partnership agreements, needs to embrace all of the main stakeholders – the local authorities, the government agencies, the voluntary sector, local business, the various local communities and certainly the anticipated funders who may include several of the foregoing, plus others.

This human and institutional spin-off from plan-making was nicely expressed by Douglas (1994: 265–6) quoting a Canadian local strategic planning task force. 'A good plan is 60 per cent process and 40 per cent content. The process – thinking, visualising, conceptualising, decision-making and problem solving – is actually more important than the plan itself...The process must provide an opportunity for people to buy in to the plan or it may just gather dust on some obscure shelf.'

As for the plan itself, it should be

- based on wide consultation and also the rigorous collection and analysis of baseline evidence, neither alone being sufficient;
- holistic and integrated in its treatment of relevant needs, issues, resources and action;
- a statement of a shared vision and a shared commitment to achieve it – a 'vision' in this context being an 'expression of what should be, as distinct from what is in place today or what might become if nothing is done' (Douglas, 1994: 250);
- selective and well-argued in its specification both of the goals and the objectives and of the measures chosen to attain them, a mere 'wish-list' being insufficient;
- clear on the partners' subsequent roles and responsibilities;
- clear on the procedures for implementation and for monitoring and evaluation; and thereby
- a genuine framework for future decision-making.

The plan must also embody the *notion of hierarchy*. This means that it should accept the need, and provide the basis, for subsequent plans containing much more detail on operational matters. Often the latter are called 'operational plans' or 'work programmes' and are shorter in their time horizon – typically one year, because of financial accounting procedures, in contrast to the multi-year perspective of a strategic plan. They spell out defined tasks each with their own earmarked resources and quantified targets as well as precise dates and the unambiguous allocation of responsibility.

This idea of hierarchy, with the strategic plan restricting itself to the 'higher ground' without being a bland concoction of worthy but anodyne statements, may be illustrated by an example relating to a hypothetical rural area.

***Vision*** 'In this area, by the year 2015, unemployment and the other major components of social disadvantage will be at a level equal to or better than the national average, and the local economy will be thriving in a manner consistent with the conservation of the area's natural and cultural environment'.
***Goals*** Eight, say, in all, including 'to empower the local community to make a significant contribution itself to the reduction of social exclusion'.
***Objectives*** 18, say, in all, including 'to provide training and other support to community groups throughout the area so that the number of people active in their communities is at least doubled'.
***Measures*** In all about 40, including 'training courses and other capacity building activities for a variety of local community activists, run in association with the local FE college and other partners'.

Normally it is at that point that the strategic plan would stop, leaving it to annual *operational plans* to set out, for example:

**Tasks,** such as 'to design, run and recruit participants for a training course for village and community hall trustees over the period October 2002 – May 2003 on "getting the most out of your hall". To be delivered mainly via distance learning and written assignments plus three weekend residential meetings.'

**Targets,** such as 'at least ten participants completing the training course'.

**Funding,** a grant of £x from the Rural Development Programme to be matched on a 50/50 basis by participants and/or by sponsorship.

**Responsibility,** the local Further Education College.

Some other 'key issues' regarding strategic planning for local development should be briefly mentioned though this is not an exhaustive list. The first is that planning is *both a technical and a political activity*. The *technical* work relates to the collection and analysis of information on the area; the execution of a programme of consultation; the elaboration of a range of possible goals, objectives and measures; the definition of possible projects and programmes; and the fashioning of appropriate methods of implementation, monitoring and evaluation. The *political* work relates to the making of choices and the setting of priorities – and to the taking of responsibility. At the heart of strategic planning for rural development, then, are exercises in bargaining, persuasion and compromise involving the different stakeholders and an acceptance that conflict, or at least disagreement, is often the norm.

It is essential, therefore, that lines of accountability are clear and that the planning process, spread over several months, is designed around a series of meetings, separated by periods of technical activity, at which agreement is reached on the main elements of the plan. In short, it is always a serious mistake if the whole planning exercise is handed over to a technical team or a firm of consultants standing apart from the political process; the result may be technically elegant but it is unlikely to be 'owned' in any fundamental sense by the partners to the exercise.

Second, and as implied in the previous paragraph, the whole process should be as inclusive as possible of the views of the various 'constituencies'. The whole of Chapters 9 and 10, relating to local partnerships and community involvement, were devoted to this theme of *actor inclusion* – though there, of course, the focus was as much on involvement in the delivery of rural development as in its planning.

Third, planning for local rural development needs to be firmly founded on a clear notion of just what 'development' is. As argued earlier, this essentially implies seeking to set in motion and to sustain a *virtuous spiral of change*, in which each action, be it focused primarily on economic, social, cultural, political or environmental concerns, is designed to be at least neutral and, as often as possible, positive in its consequences for the other dimensions. Unfortunately, strategic planning documents are often couched instead in the language of 'balance' – meaning 'a bit more of this and a bit less of that' – and not in that of a developmental 'spiral'.

Fourth, and related to this, is the temptation to squander scarce resources on *gap-filling*, meaning attempts to correct deficiencies in provision by the mainstream agencies. For example, if the baseline study reveals significant localities lacking a daily bus service or having insufficient units of affordable housing, then the temptation is to put right the deficiency by writing appropriate measures into the strategic plan. But this is ultimately a battle that cannot be won because there will always be gaps in provision and always people clamouring to have them filled. This is not to argue that those gaps do not cause hardship or that the claims are necessarily invalid, merely that a strategy for local development must focus on just that – on how development, the 'virtuous spiral' referred to above, can best be initiated and supported with the scarce resources available.

## THE TOOLKIT

We now turn to the tools available to undertake the planning exercise. There are several stages involved, though much of the work can be carried out simultaneously. Time is often tight and if momentum is not to be lost, six months may well be the maximum time available to move from a standing start to an agreed and published plan on which all the main actors are agreed. And, of course, that period has to embrace both the technical and the political process, as explained above.

There is not a universally accepted model of strategic planning, but most commentators, certainly those cited above, typically indicate *five essential stages,* and these are now considered in turn with local rural development in mind:

### 1. CLARIFYING THE CONTEXT

This is the 'diagnostic' stage. Chapter 11 was wholly devoted to this subject; suffice it to say here that its real purpose is to define the key concerns that demand attention in the strategic plan and the context within which practical measures may be fashioned.

### 2. AGREEING A VISION, GOALS AND OBJECTIVES

Remember that those three terms are employed in a hierarchical sense. In practice, the *vision* tends to emerge as the diagnostic work is undertaken and reflected upon; the task is often more that of crystallising it in clear, concise language upon which all, or most, can agree. But specific tools are available to aid 'collective visioning', such as the organisation at a very local level of 'future needs days'. These involve as many people as possible in a series of practical tasks, using models, maps, pictures, flipcharts and the like to help them brainstorm and then reflect upon their likes, dislikes, hopes and fears for the locality as a whole and for specific 'hot spots and grot spots' within it (see Northamptonshire ACRE, 1999 for a good exposition of this tool in a village context).

The *goals and objectives* should flow from the vision and again be firmly based on the evidence of the diagnostic work. They should be *comprehensive*, spanning the domains in which action is intended, *realistic*, not making unreasonable expectations about the resources likely to be available, and sufficiently *clear* that they can form the basis of the monitoring and evaluation that will follow later. But most of all they should set out what it is intended to achieve.

Inevitably, choices have to be made in agreeing goals and objectives. All local rural development programmes have to prioritise between the various issues, localities and social groups that will clamour for attention. In the case of issues, is personal accessibility or affordable housing or environmental protection or job creation or agricultural diversification, to be given priority? In the case of localities, are the small towns, say, or the run-down former coal mining area or the remote villages to be given priority? And with regard to social groups, are the long-term unemployed or elderly people living alone to be given priority or, say, small farmers in difficulties?

The agreed goals and objectives should reflect decisions on such questions, but how are they to be made? Suggestions may come from a sample social survey or by focus groups, whereby the residents of the area are asked to rank their levels of concern on a number of issues. Some sort of 'gaming exercise' involving a quasi-voting procedure can help to sharpen the choices and trade-offs in this context. In addition, some technical analysis, seeking, for example, to quantify both the 'need' and the 'potential' of different localities in the area, using several statistical measures in each case, can provide a way of exploring where resources might be channelled to best effect.

But really all such exercises can only be the prelude to political choice. In that respect the challenge is to engineer a discussion and decision-making process which involves all the various stakeholders and which gives each a chance to argue their views, and in this a skilled facilitator can be invaluable. The main thing is to avoid either conceiving goal-setting as a wholly technical process for the experts or leaving it to small cabals of political heavyweights – the 'smoke-filled room' approach to priority setting.

### 3. DECIDING IN OUTLINE FORM THE MAIN MEASURES OR ACTIONS

The next question concerns how the goals and objectives are actually to be realised given the resources available and the constraints at work. In this respect, the relevant 'measures' or 'broad actions' will almost certainly have begun to suggest themselves during the two preceding stages. But it is valuable at this stage to begin with a few fundamental questions. One such question is, 'Will we rely on funding a number of discrete projects to attain our objectives or is there scope for putting some resources into seeking to influence mainstream programmes run by one or more of the key agencies?' For example, it may not be sensible to put £30,000 into three discrete training initiatives and ignore the £3 million budget of the area's main training agency – in effect leaving the latter to go on deciding its own priorities and actions oblivious to the agreed rural development programme. This is where a well-constituted local partnership can prove invaluable, with the representative from the 'training agency' encouraged to champion the goals of the agreed development programme in the agency's day-to-day work.

Responding to questions such as that will help to define the sorts of measures that are to be written into the strategy. But in the end, the task is one of scanning the statements of issues, needs, resources, opportunities, goals, objectives etc., and asking, in relation to each, what are the policy or programme options, and what sorts of measures have other agencies or other rural development programmes promoted in similar contexts? As ideas for action begin to emerge, they will need to be subjected to at least an informal assessment regarding both their feasibility and their capacity for making an impact on the issue in hand. But, as explained earlier, most strategic planning exercises will not go beyond that level of specificity. It is the following stage, operational planning, that rigorous project assessment comes in.

### 4. ESTABLISHING PROCEDURES FOR IMPLEMENTATION

Chapter 13 will consider 'implementation' in greater depth. Here we will simply set out some of the implementation issues that need to be resolved at the *strategic planning stage* and clearly expressed in the plan. They include:

- How will annual 'operational plans' or 'work programmes' be drawn up and approved?
- How will they be implemented and how will any slippage and associated underspend be dealt with?
- What will be the balance between *pro*active projects – those designed and executed by the development partnership itself – and *re*active projects – those designed by third parties and submitted for approval and funding?
- How will project proposals be solicited and assessed and, if successful, supported and monitored? What will be the characteristics of a 'good project'?
- What will be the expectations about 'matching finance' – i.e. the 'topping up' from other sources of the funding to be made available via the programme itself?
- What will be the in-house decision-making structure and procedures? For example, how will responsibilities be distributed between staff, the 'board' and any sub-committees? How often will the committees meet? What will be expected of individual partners between meetings? How will financial management be handled?

### 5. ESTABLISHING PROCEDURES FOR MONITORING, EVALUATION AND THE UPDATING OF THE PLAN

Chapter 14 will consider monitoring and evaluation in detail. Suffice it to say here that those activities are necessary to check the plan's progress, continuing relevance and effectiveness in achieving the desired results.

And the plan, once approved, should not be considered to be unchangeable; far from it; if circumstances change in a fundamental way, so must the plan. Generally, however, strategic plans for local development are written in a 'robust' manner – in a way that renders them likely to remain a valuable framework for decision-making for at least five, say, of the typical 10- to 15-year period in prospect. Hence the notion of 'hierarchy' and the need to relegate a lot of detail to annual work programmes or operational plans. But after a while the plan will be 'showing its age' and will need to be revisited and revised by going through some at least of the steps set out above. There must be no sense of 'failure' if a strategy with a 15-year perspective is substantially rewritten after five – such is the nature of a 'rolling programme', meaning a succession of long-term plans, periodically revised before they expire.

Many of these principles and tasks are demonstrated in two case studies that relate, respectively, to a quite extensive rural area and to the village or parish level.

## CASE STUDY 23: THE DORSET RURAL DEVELOPMENT STRATEGY 1994–98

In December 1993 a large part of rural Dorset (population 136,000) was designated a Rural Development Area by England's Rural Development Commission, with the promise that it would part-fund, as part of a local partnership, a Rural Development Programme (RDP) aimed at alleviating some of the area's social and economic problems and at setting in train a wide-ranging development process. By early 1994 a Committee for Rural Dorset had been established, comprising 15 local or regional partner agencies to manage the programme, and by May 1994 a Dorset Rural Development Strategy had been drafted and agreed. It is that document, the May 1994 Strategy, comprising 33 pages of text plus a number of maps, that we will focus on here (Committee for Rural Dorset, 1994).

It announced (p. 7) that

> an essential part of the Rural Development Programme process [is] the production of a strategy that identifies needs, opportunities and resources within rural Dorset, and sets objectives and a framework for action. The strategy should look ahead over a ten year period, but concentrate mainly on the first five years. It will be subject to a full review after five years but should be revisited each year as part of the process of updating the 'operating plan'. The operating plan is the means for translating the strategy into action, via a three year rolling plan of action.

The strategy itself, as set out in subsequent pages of the document, followed quite closely the 'model' set out above. First was a straightforward *profile* of the area covering a number of familiar elements, notably population trends, housing, car ownership, employment/unemployment, tourism, rural facilities and environment. Second, and linked to that profile, was the crystallisation of about 40 *key issues* which 'will provide the context for the development and prioritisation of strategic aims and objectives'; for example, 'How can the current imbalance in age structure in the rural communities be most effectively addressed?' and 'How can the area's environmental quality be enhanced to improve the county's appeal to tourists?' This listing of issues was set within a text which explained and elaborated on their importance.

A shared *vision* was then set out, derived both from the aspirations of the Rural Development Commission and a discussion involving all the other partners. It comprised 'balanced social communities living a sustainable lifestyle, with good access to a range of employment and training opportunities, the local housing market, a choice of community services and facilities, and able to appreciate and enjoy a high quality of natural and built environment'. It continued, 'this vision may seem to be ambitious and a long way from our present situation, [but] the purpose of this strategy is to...set a course towards achieving the vision' (p. 26).

The 'vision' was followed by a *strategic aim* similarly referring to the resolution of social, economic and environmental problems through the development of a sustainable rural economy, and then, more pointedly, by eleven *objectives* which are set out below in a somewhat abbreviated form:

1. to secure a more diverse range of employment opportunities;
2. to encourage and facilitate enterprise by local people and businesses;
3. to ensure the provision of an adequate supply of land and buildings for employment purposes;
4. to increase the scope, quality and flexibility of training provision;
5. to assist the revitalisation of the area's tourist industry;
6. to assist the provision of an adequate supply of affordable/social housing;
7. to provide and maintain a range of community facilities and services;
8. to support an infrastructure ... which enables local people and groups to articulate their views ... and facilitates local action to meet local needs;
9. to promote the development of public transport services and alternative transport modes;
10. to promote respect and enhancement of the natural and built environment; and
11. to undertake survey/research work ... to increase understanding of rural communities, lifestyles and trends in order to inform and/or to lobby.

A final, brief section of the document dealt with *implementation and evaluation*. It stressed the need for co-operation and co-ordination, both between the partner agencies and local voluntary and community organisations and, in a different sense, between the Rural Development Strategy and all the other strategies adopted by the 'mainstream agencies' dealing with land-use planning, the countryside and tourism, for example. It also explains that, each October, an 'operating plan' would be prepared for submission to the Rural Development Commission, detailing the projects for implementation the following year and requesting appropriate funding for them. Very little was said about 'evaluation' except that progress was to be reviewed each year and that the 2001 census would eventually allow an assessment of progress in the Rural Development Area compared with that in neighbouring rural areas.

Impressively, the Strategy document did remain a genuine basis for decision- making, at least until 1998 when the present author was commissioned to undertake a mid-term appraisal of progress over the preceding four years (Moseley, 1998). Each annual *operating plan* comprised two main sections: the first reviewed progress in the previous year, both in relation to the dozen objectives set out in the Strategy, and with regard to each specific project that had been specified in the previous year's plan; the second section set out in tabular form the projects, usually about

50 or 60 in number, which were in prospect for the forthcoming year – each one cross-referred to the relevant objective(s) and with an associated funding stream set out for the coming year and, where appropriate, the two subsequent years as well. Thus these 'operational plans' were firm and precise for one year ahead and indicative thereafter.

The Dorset RDP Strategy, therefore, mirrored quite closely the model of strategic planning described earlier in this chapter. In that respect it typified quite well the 30 or so RDP strategies produced for the various English Rural Development Areas in the 1990s. The strategy set out clear objectives and, via the Operating Plans, ensured a reasonable fit between the individual projects that were subsequently funded and the strategic goals of the whole exercise.

But looking simply at the Strategy document itself, it was a little disappointing in several respects. Recalling the simple criteria suggested earlier for appraising the quality of such documents, it was clearly much better at outlining 'what we want to do' than 'how we are going to do it' – perhaps because of an implicit assumption from the outset that the method was going to be nothing more than 'a call for good projects' that would demonstrably help achieve one or more of the objectives. Nor did it say much about 'how will we know if we are doing it right?' though the Operating Plans subsequently fulfilled a valuable monitoring role.

In addition, there is no indication in the document of the *process* whereby the strategy was generated and approved. It simply, as it were, 'falls from the sky'. Who, one wonders, was really behind its formulation and how was consensus reached? But the interweaving of the technical and political processes underlying it is not explained. As a corollary of this, there is no sense that the strategic planning exercise was itself a 'capacity building' or mutual learning exercise in its own right; indeed, there is no indication that any consultation occurred beyond the small circle of partners.

Happily, the author's own evaluation of the Dorset RDP four years on indicated that there had by then been a good deal of 'mutual learning', especially involving the middle-ranking officers in the main eight or nine agencies active in the partnership. And the spin-off or synergy resulting from the business of managing and implementing – if not initially devising – the strategy was a real and significant benefit of the whole exercise.

---

## CASE STUDY 24: VILLAGE ACTION PLANNING AND PLANS

---

In recent years an increasing number of 'local communities' – meaning groups of people concerned for the future of their village, parish or small town – have worked to develop and implement some sort of 'strategic plan' at the very local level. Often this has been a wholly informal exercise with the parish council and higher-level statutory bodies being involved only tangentially, if at all. This burgeoning of very local planning reflects several features of England in the late 1990s – notably a reduced readiness on the part of local communities to accept what is often adverse change emanating from outside, a growing awareness that the pursuit of 'sustainability' implies community empowerment as much as environmental protection, a shift in central and local government policy towards 'enabling' the provision of rural services rather than their top-down provision and the sheer availability of funding and encouragement to 'do it yourself' coming from LEADER groups, the lottery and government or local authority small-grant schemes.

In 2001 the author was commissioned by the Countryside Agency to take stock of these very local strategic planning ventures, concentrating on six English counties and 13 localities with between 200 and 6000 inhabitants (Moseley, 2002). The focus was to be 'attempts by local rural communities in England to produce and implement holistic action plans for their immediate localities'. To be 'in the frame', the plans had to be just that – 'plans' and not just community appraisals – and also quite wide-ranging rather than devoted to one specific issue. Thus defined, there seemed to be scores rather than hundreds of 'village action plans' nationwide in 2001, though with many more in the pipeline.

The best all had six key characteristics:

- *holistic in scope* – at least trying to treat the issues of concern to local communities in the round rather than in separate compartments;
- *bottom-up in approach* – with the genuine and thorough involvement of the local community in the planning exercise, and some measure of local ownership;
- *research-based* – in the sense that an effort was made to gather and appraise evidence on local issues and resources as well as on local needs and hopes for the future;
- *concerned with priorities* – involving an attempt to reconcile local conflicts between objectives and disagreements over priorities;
- *wedded to a partnership approach* – bringing together in the implementation phase both the local community and official agencies and organisations; and
- *focused on action* – with a clear elaboration of the intended projects and of the why? by whom? where? when? how? and with what resources? aspects of all proposed action.

At least half of the 13 local communities in the six counties studied in-depth had effectively followed an eight- or nine-point process, advised throughout by a local development agency of some kind, normally a Rural Community Council or local authority:

1. an initial awakening of interest in the parish;
2. a public meeting at which the idea of an action plan was aired;
3. the formation of a 'village action group' at first just to plan a village appraisal;
4. the execution of the village appraisal and possibly of other 'audit' exercises;
5. the presentation of the appraisal report at a second public meeting;
6. the preparation by the action group and sub-groups of a draft village action plan based on the appraisal;
7. the presentation of the action plan to a third public meeting for discussion and ratification;
8. attempts at implementation by individuals, small groups or agencies; and
9. (sometimes) the monitoring and evaluation of progress.

Various conclusions may be drawn from the 13 case studies about the state of 'village action planning' in 2001. In particular:

1. The support of a *local development agency* was crucial – offering encouragement, technical advice, modest financial support and routes to funding sources.
2. In all cases, some sort of *local action group* provided the driving force. They tended to be most effective when they federated small working groups and when they adopted an open stance to the recruitment of new members and to their relationship with the rest of the community.
3. *The parish and town councils* played a variety of roles. Some initiated the planning exercise and remained strongly involved throughout. Some were supportive but remained at arm's length, receiving and commenting on papers and enjoying some cross-membership with the action group. Others remained on the sidelines, not always by choice.
4. *Social inclusion* was everywhere extolled as an ideal but 'exclusion' inadvertently crept into the planning process at several stages;
5. Village appraisals comprised the most common form of *initial baseline study* but they were deficient in a number of respects as a consultation tool and they placed excessive weight on identifying local *need* as distinct from *resources*.
6. Actual *plan-making*, as distinct from 'fact-finding', was generally treated as an informal, creative process; the plans tended just to 'emerge', albeit with some degree of subsequent local validation.
7. Often little explicit attention was paid to the creation and support of *machinery for implementing* the projects and proposals that emerged from the planning exercise.
8. The evidence indicated that *a great deal of value* normally flowed from these village action planning exercises – both projects on the ground and the 'soft outputs' of enhanced 'capacity' and of community development.
9. These practical outputs depended largely on two essential ingredients:

- the quality, energy and tenacity of the local volunteer resource and the quality of the planning and implementation work undertaken by these volunteers; and
- the quantity and quality of the support that these volunteers received from a local development agency.

By way of a specific example we may cite the Staffordshire parish of Ipstones (population about 1450). In 1997, and again in 2000, an Ipstones 'parish action group' with a dozen or so members including some parish councillors, produced a 'parish action plan' which was effectively legitimated by well-attended public meetings. The first plan was linked to the Peak District's LEADER II programme, the second to the subsequent Staffordshire Moorlands Single Regeneration Budget (SRB) development programme. Significantly, in each case there was a requirement that local communities wanting to access project funding had first to demonstrate a high level of local support for the alleged priorities, by means of a thorough action-planning exercise.

The production of the two plans closely followed the 'nine-point model' set out above. Each was based on a parish appraisal involving a household survey and upon public meetings at which voting on possible projects took place. Sub-groups of the action group played an important role in subsequently working up project funding applications and managing the implementation of the various projects. There were some significant successes, including an annual Christmas lights festival which brought substantial business to local traders, the creation of four units of sheltered housing in a redundant school and the construction of purpose-built accommodation for children's play activities. And everyone agreed that local community spirit and the pool of volunteers prepared to 'have a go' did increase as a direct consequence of the planning exercises and the subsequent action on the ground.

Looking at the flurry of activity that occurred over the five years in question, it is clear that the substantial support provided by the LEADER and SRB fieldworkers was crucial; indeed the SRB programme now deploys part-time 'village agents' to offer support in just three or four parishes at any one time. It is also clear that Ipstones' two parish action plans each blurred the distinction between 'strategic' and 'operational' planning. They were, in part, a programme to be systematically implemented and, in part, a trigger for ongoing local creativity and initiative. And interestingly, the research concluded that 'in Ipstones it was the planning process rather than the written plan that has been the locomotive of local development' (Moseley, 2002).

Thus, in neither of our two case studies was the classic strategic planning model outlined earlier followed precisely, but each does show the value of systematic thought and action adapted to local circumstances, and each successfully orchestrated concerted action.

---

### SEE ALSO...

Case Study 13 also relates to the Dorset Rural Development Programme, but specifically from a 'social inclusion' viewpoint. Case Studies 17 and 20 provide other examples of strategic planning for local rural development.

### SELECTED FURTHER READING

On local area strategic planning see Douglas (1994), Area Development Management (1994), Flanagan *et al.* (1995) and LEADER Observatory (1995b). On such planning at the village level see Moseley (2002).

# 13

# Implementation: Making Things Happen

> 'Praise the Lord and pass the ammunition.'
>
> Howell Forgy, American naval chaplain at Pearl Harbor, 1941

## DEFINITIONS AND IMPORTANCE

Reflecting on his experience in promoting local development in Canada, Douglas (1994: 255) concluded that 'the Achilles heel of planning, particularly in the public, quasi-public and community arenas, is the yawning gap between so many plans as designs, and plans as operational management vehicles. Little wonder that too many plans fill too many dusty shelves.' It is with the challenge of narrowing that 'yawning gap' and of ensuring that development actually happens on the ground that this chapter is concerned – a challenge insufficiently aired either in the academic literature on rural development or by those who actually devise development strategies at the local level. For every dozen treatises on 'planning' or on 'strategy' there is barely one on the practicalities of 'implementation' (but see Flanagan *et al.*, 1995; LEADER Observatory, 1995b and 1998).

We will consider 'implementation' primarily as it relates to local area development programmes, which themselves generally contain or imply scores of individual activities or tasks to be accomplished over a number of years. The previous chapter introduced the need for this by explaining how long-term and necessarily wide-ranging strategic plans 'hand over' to short-term and more precise 'operational plans' designed to put them into effect. It stressed the role of individual 'tasks' linked to strategic objectives, each with their own targets, costing and allocation of responsibility.

That ground will not be covered again here; rather we will concentrate on the nature of 'projects' and their selection, development, support and monitoring, as well as certain issues relating to decision-making and the necessary management structures and procedures of the development agency. Other chapters are also relevant in this respect, of course, notably those concerned with the role of local development partnerships, with community involvement and the promotion of entrepreneurship – each of them addressing the challenge of mobilising the various actors who are crucial to the development process. The point is that implementation does not somehow automatically happen once a clear strategy and a succession of operational plans have been agreed; it is not just a matter of systematically carrying out a predetermined plan. Implementation must be creative and developmental itself, fostering initiative and enterprise as projects, set within the plan, are solicited or devised, worked up into operational form and subsequently supported through to their conclusion.

As for a definition, it is suggested that *implementation* is the process of putting into effect and successfully accomplishing the measures and tasks contained in strategic and operational plans.

## SOME KEY ISSUES

Many 'local development agencies' seem automatically to assume that having devised their strategic plan, the challenge of implementation consists only of soliciting appropriate project proposals from people and businesses in the area, choosing which ones to support and putting them into effect by offering funding and other practical assistance. Much of implementation is indeed about this, but there are two fundamental questions to address first.

### THE 'INFLUENCING ROLE'

The first question concerns the balance to be struck by a development agency between, on the one hand, promoting discrete projects and, on the other, seeking to influence other actors and organisations whose work is relevant to the development agency's own aims. This 'influencing role' has often been neglected but can be very valuable, especially as the 'other actors and organisations' may well have substantially more resources at their disposal. Persuading a large health authority to switch one half of 1 per cent of its resources towards a certain area or target population group may well do more for the disadvantaged residents of the area in question than spending, say, 25 per cent of the development agency's own budget on a specific healthcare project.

This is increasingly recognised by the British government, especially in its insistence that local authorities now accept a general responsibility for the well-being of the residents they serve and do so, in part, by devising and implementing area-wide 'community strategies' in close association with other local agencies (Department of the Environment, Transport and the Regions, 2001). Facilitation and influencing are key elements of this. In similar vein, 'voluntary sector compacts' – agreements between public and voluntary sector agencies in an area – set out how the agencies will work in partnership to nurture community activity and generally improve the quality of life of local people. Again, 'influencing' is a key element of implementation.

Hence the value of local development partnerships, discussed more fully in Chapter 9, of 'signposting' to other sources of help and of actively building and using networks. Indeed, 'networking' – meaning the systematic cultivation and nurturing of 'contacts' or individuals who are influential either in their own right or by dint of their position in other organisations – has emerged as a major tool of implementation in recent years. The task is to keep such people informed of your concerns and activities, to seek to influence their activities and to solicit their support for yours. In practice this means devoting resources to appropriate communication media – websites, newsletters, seminars, exhibitions and the like – to building up some sort of membership base of people with shared concerns, to lobbying and persuasion, to partnership working and to community involvement. (The 'Peak District Rural Deprivation Forum', described in Case Study 14, provides an excellent example of effective 'networking' as a deliberate component of a local development programme.) It would be an interesting research exercise to take a sample of recent rural local development programmes and to assess what proportion of their eventual success arose from 'influencing others' as distinct from their direct support of funded projects.

### PROACTIVE AND/OR REACTIVE?

A second key issue for a local development agency, as it considers how to implement its strategic plan, concerns the proper balance between proactive and reactive projects. The former are those devised and carried out by or for the development agency itself; the latter are those which emanate from third parties, be they private individuals, companies, co-operatives or community groups, and are brought to the agency with a request for funding and support. Sometimes proactive projects play no part in a development programme, in which case the agency is exclusively, and perhaps vulnerably, dependent on what happens to come forward for its consideration. At the other extreme, proactive projects can so dominate a programme that the 'bottom-up' conception of development is virtually forgotten.

Generally a balance is preferable, with the soliciting and support of bottom-up projects warranting priority if the basic infrastructure of development is already in place and a culture of enterprise exists. Conversely, in LEADER areas 'where preliminary co-ordination and mobilisation work is necessary to "flush out" potential project leaders', it has been suggested that over half of the local programme's budget might usefully be devoted to proactive projects (LEADER Observatory, 1998: 9). LEADER groups in that position have felt obliged to invest heavily in various awareness-raising initiatives, training programmes, demonstration projects and the deployment of project officers with a brief to go round and 'stir things up'.

Many have also invested in 'flagship projects', involving the local action group in creating some substantial and very visible facility around which private project promoters might devise associated and generally smaller projects. An example is the renovation by a French LEADER group of a number of long-neglected castles in the Pays Cathares as a way of promoting the identity of the whole area and of stimulating a mass of small heritage tourism initiatives in the vicinity (LEADER Observatory undated b).

## THE TOOLKIT

### *PROJECT SUPPORT*

We now go on to assume that those two issues have been resolved – namely the balance to be struck between exerting influence and funding projects, and that between supporting proactive and reactive projects. We will also assume that the development agency has both a clear strategic plan and a budget to part-fund a number of reactive, bottom-up projects, coming forward from local 'project promoters' with requests for support. In that scenario, the author's review of the effectiveness of rural development partnerships suggested that 'the model rural development partnership should be user-friendly to existing and potential beneficiaries of the partnership's support through providing accessible information, easily available advice and support, easy-to-use funding processes and good follow-up' (Cherrett and Moseley, 2001: 13). This prescription for good practice in local development is now explored further with reference to the soliciting, selection, funding, support and monitoring of projects.

#### *1. The soliciting and 'working up' of project proposals*

It is rarely enough simply to announce that the agency or partnership is 'open for business' and then wait for people with well-founded project proposals to come forward. The first step is generally the production of some sort of easily understood 'prospectus' spelling out what the development programme is about and explaining both the sort of projects that might be eligible for support and the procedures to follow. This 'prospectus' has then to be promoted and explained using many of the techniques of community involvement discussed in Chapter 10. Chambers of commerce, town councils, community groups, groups of farmers and other local businesspeople need to be approached and their interest secured. Some basic 'capacity building' work may be necessary if the level of entrepreneurship in the area is low. Indeed, the task may well be to help potential project promoters develop coherent and well-argued project proposals from what at first may be only very sketchy ideas. All of this takes time and diligent work by 'project officers' and other ambassadors of the development programme, and a readiness to work with project promoters and possibly other stakeholders sharing an interest in the project's success.

### *2. The selection of projects*

Assuming that a steady stream of project proposals then comes forward, all with some sort of outline 'business plan' making clear their purpose, key characteristics, proposed management and likely outcomes, what should happen next? Generally, a selection process ensues, the key element of which is the careful scrutiny of the proposal against a list of criteria. Case Study 25 presents such a list of criteria as used by a Scottish LEADER II group, but essentially three judgements have to be made. The first concerns *congruence with the strategy* – How well does the proposed project fit with the agency's objectives? Does it promise to meet identified needs and to contribute significantly to the area's development process? Second, regarding *the project itself*, relevant questions might include: Will it really 'bite' on those needs? Is it realistically costed and financially viable? Is there a market niche for it? Will it offer synergy with other projects? Third, some questions about the *project promoter and/or the proposed management*: Will the project be soundly managed? Do its promoters have the necessary experience or a determination to learn? If a partnership is involved, will it gel and add value to the enterprise? Positive answers to such questions will be needed, while always remembering that some risks must be taken in what is an entrepreneurial situation; the best rural development programmes are often those whose projects enjoy a 90, rather than 100 per cent, success rate. An absence of any failed projects over a period of years may mean that the development agency has been insufficiently adventurous.

### *3. The funding of projects*

Generally, a development agency will offer to meet only a relatively small percentage of the start-up and/or running costs of a project, the actual amount varying with a number of factors such as the scale and type of project and the degree to which it meets some of the criteria referred to above. In any event, 'co-finance' will be needed from other sources which, depending on the nature of the venture, are likely to include the applicant him/her/itself, commercial banks, local authorities and charitable sources. The importance of co-finance is not only that it allows the development agency's own resources to be spread further; it also provides the reassurance that others are persuaded of the project's merits. Sometimes a significant input of voluntary labour will be accepted in lieu of co-finance and this can be a crucial concession if voluntary sector projects are to be brought forward.

A host of other funding-related issues affect the viability of projects and hence the implementation of rural development programmes. They include problems of cashflow, when funding is paid late or in arrears, and the general complexity of the financial procedures to be followed. (In order to be able to secure lottery money for major refurbishment projects, some English village hall committees, for example, now have to commission expensive feasibility studies and make complex parallel applications to other funders with different criteria and procedures.) 'Delegated grant schemes' are often valuable in this respect; these involve some intermediate body such as a local development trust or voluntary organisation receiving one large grant from a funding agency and then disbursing it to very small projects with a minimum of fuss.

***4. The support of projects***

Many of the projects that are considered by LEADER groups or other development agencies come from people or local groups with only limited experience of running a business or community development project. They may need training, help with securing further finance, business advice or technical assistance of various kinds as the project gets underway. As consultants reviewing the progress of LEADER II programmes in Eastern England remarked, 'project officers or animateurs have played a vital role in helping to develop projects with community groups and other applicants. Their role has often involved not just assisting with funding applications and reviewing business plans but also a good deal of "hand holding" in maturing embryonic projects' (Segal Quince Wicksteed, 1999: 8). This assistance need not necessarily come from salaried project officers. Often experienced, perhaps recently retired volunteers with a relevant background can be of great value in passing on their expertise to inexperienced entrepreneurs. Indeed, for some such people joining the management board of a LEADER group, for example, it is this 'help on the ground' role that they really enjoy, rather than that of discussing strategy at a series of committee meetings.

***5. The monitoring of projects***

Finally, there is a need for the development agency to monitor progress, 'monitoring' being 'the discipline of assessing en route the progress of the plan and of its components as a basis for identifying remedial action' (Douglas, 1994: 260). (It is not to be confused with 'evaluation' – the more fundamental scrutiny of the performance of a project or programme, reviewed in the next chapter.) Procedures to gather the relevant information need to be put in place and linked to appropriate indicators and 'milestones' which are key stages that have to be reached before further progress is possible, and a timetable has to be set for periodic reflection on the information assembled.

In monitoring, attention must be paid both to:

- *individual projects* Are they proceeding satisfactorily? Are their output targets being met? Are problems being encountered? For example, 'Has the new bed and breakfast accommodation been provided, with the hoped-for 50 per cent occupancy rate?'

- *the development programme as a whole* For example, 'Are we on track to spend all of this year's allocated budget?' 'Are we achieving our target of giving decisions on project applications within six weeks?'

Unfortunately, the monitoring of a project's outputs can become just a mechanistic and wearisome exercise undertaken just to keep the project's sponsors happy. This is particularly likely to happen when the promised 'outputs', reluctantly signed up to as a condition of funding, in fact bear little real resemblance to what the project is really about. Thus an innovative, locally delivered programme of support to community enterprises across the rural south-west of England, funded under the government's Single Regeneration Budget (SRB) programme, was seriously distracted by a requirement that every month the various participants quantify their performance in relation to 12 output measures taken from a standard SRB 'menu', 11 of which seemed to the author to bear little relevance to the task in hand (Moseley, 2000b; see also Case Study 9). Examples of measures which seemed irrelevant and of dubious validity and practical value were 'the number of voluntary groups supported, the number of training weeks delivered, the number of local firms with volunteering schemes [and] the number of young people developing their social and personal skills'. Only one of these related to 'community enterprise' which was what the programme was about, and much of the time of the project's periodic team meetings was generally devoted to the refinement of data for inclusion in a vast matrix of output measures. It was no wonder that one project manager confided, 'We all hate this silly numbers game – it doesn't reflect what we are doing.'

In short, monitoring is useful, but the lesson is 'Beware of just going through the motions.'

### SUCCESS AND FAILURE

Almost without exception, the various evaluations undertaken of the performance of the UK's LEADER I and II programmes stress the crucial role of the 'project officer' or 'animateur'. Sometimes these people are centrally managed, working from the LEADER offices. Sometimes, particularly where very large, sparsely populated areas are involved as in the Scottish Highlands, local part-time agents may be recruited in far-flung corners of the area in question, with much the same role. Their duties would include bringing people together, soliciting and developing project proposals, helping them write 'business plans', guiding project promoters through the financial and bureaucratic maze, taking the proposals to committee and undertaking the subsequent communication, follow-up and monitoring. Among the attributes they need are energy, commitment, flexibility, networking skills, administrative competence and, of course, a grasp of what exactly is 'local development'. In its mid-term evaluation of England's whole LEADER II programme, the consultants concluded that 'the resources spent on project officers/animateurs (by LEADER II groups) has been money well spent. It is inconceivable that any significant progress could have been made without these individuals'. (PIEDA Consulting, 1997: 58).

Other factors that have helped to determine the degree of success enjoyed by local development programmes run along LEADER lines include:

- enjoying good relations with both the local authorities serving the area and the major programme funders;
- having a well-balanced and effective 'local development partnership' (see Chapter 9 on this point); and
- enjoying a 'running start' in the sense of being able to build upon previous endeavours to promote local development in the area which will have done some of the groundwork – raising awareness, developing entrepreneurship, building networks etc.

Conversely, the same mid-term review of England's LEADER II programme concluded that

> 'overall, progress in terms of project and fund approval and outputs achieved has been slow in most areas ... there are four main explanations:
>
> - delays in the implementation of the national programme (i.e. in the setting up of the local action groups and of the regional machinery for overseeing them and for channelling project funds via them)
> - an underestimate of the time required to build capacity and form partnerships within rural communities
> - the complex and bureaucratic nature of the project application process beyond the local level
> - inadequate resources devoted to project development and the project application process at local level in some areas.' (*ibid.*: 70)

Happily the various regional evaluations of the LEADER programme that reconsidered the position two years later (e.g. Segal Quince Wicksteed, 1999; Atlantic Consultants, 1999) found that by then much more substantial progress had been made – evidence once again that 'local development takes time'.

### CASE STUDY 25: THE SELECTION OF PROJECTS: SOME LEADER II EXPERIENCE

---

**The Oswestry Hills LEADER II programme** was one of the smallest in England with about £1 million to distribute to appropriate projects over about five years. It was managed by a LEADER Action Group (LAG) of 18 organisations representing the main funding partners, the local authorities and various local business and community organisations. One full-time and one part-time project officer were employed.

According to UK LEADER II Network (2000), the LAG generally distributed funding to project applicants as follows:

1. Applicants approached the project officer with an idea and, with his/her assistance, developed it into an eligible project proposal and set out the details on the LEADER application form.
2. The project officer then scored applications against agreed criteria and presented them to the LAG at one of its bi-monthly meetings, the applications being approved, rejected or referred back for more information.
3. Successful applicants were then informed that they had next to get approval from the regional Government Office or the Ministry of Agriculture following an 'eligibility check' based on a 'Structural Fund application form', which the applicants had to fill in.
4. They were also offered the option of 'bankrolling' by the Shropshire Chamber TEC, in recognition of the cashflow problems that might arise from the payment of grant in arrears. For this facility a cashflow analysis was required.

Stages 1 to 4 could, and usually did, take several months to negotiate even for projects needing just a few thousand pounds. And it should be remembered that LEADER would only fund up to 50 per cent of a project's costs, so parallel procedures, often operating to a different timetable and with different application forms, had also to be negotiated with other funding bodies if the applicant was unable to find the 'co-finance' him/herself. Little wonder that the mid-term evaluation of England's LEADER II programme reported that 'the large amount of bureaucracy involved in accessing what amount to small grants is very frustrating for everyone involved' (PIEDA Consulting, 1997, appendix C: 1) – a conclusion by no means applying just to the Oswestry Hills experience.

Despite all this – and the subsequent bureaucracy which involved all project promoters completing detailed quarterly returns detailing their 'outputs' as well as financial expenditure – many good projects flourished. These included training courses for volunteers, a family centre providing childcare where none existed before, a youth cyber-café and a 'community action grants' project which involved the local authority distributing very small grants (usually of a few hundred pounds) to community groups, thereby sheltering them from the bureaucracy described above. In all of this, the skill, dedication and diligence of the project officers were absolutely vital.

As for *the actual criteria* used by LAGs to help them decide which project applications to approve or reject, the experience of the **Scottish LEADER II group 'Rural Inverness and Nairn'** (*www.leader2.org.uk*) may be usefully described.

There were nine *mandatory* criteria. In order to be funded, a project had to be

- consistent with one of the LEADER II business categories – for example training, rural tourism, the small-firm and craft enterprise sector, the exploitation of agricultural products;
- innovative in process, product or market as far as the local area was concerned;
- designed to deliver tangible community benefits such as new jobs, increased human 'capacity' or other socio-economic (not just environmental) outputs;
- 'additional', in the sense that LEADER II funding was essential for the project to succeed and to produce a 'net gain' for the local community;
- capable of serving as a 'demonstration' project which would encourage other possible promoters;
- sustainable in terms both of sensitive resource use and of economic viability;
- able to demonstrate a demand or market;
- able to demonstrate a lack of 'displacement influences' (meaning the mere replacement of existing activity in the area); and
- demonstrably good value for money.

In addition, preference was given to proposals which

- formed part of a locally designed community development strategy or appraisal;
- stemmed from a legally constituted community group;
- could demonstrate good community support;
- could lever more than 30 per cent of the total project cost from community or private investment;
- applied to more than one of the LEADER II business categories referred to above; and
- favoured one or more target groups such as the local unemployed, women returners, young people and diversifying farmers.

That is a tough list of criteria, and it is not known whether they were all afforded equal weight and how far, in practice, successful projects met them. Clearly, such lists – which were widely used by LEADER II groups and were by no means identical from area to area – posed at least two management challenges. First, how to determine or measure compliance (how does one actually recognise an 'innovation' or a 'sustainable' project or 'good value for money', especially in advance of the project's launch?). Second, how far to overlook a proposed project's deficiencies if the local action group – a collection of individuals brought together to use their judgement and experience and not just to tick boxes – still felt that it was worth backing?

In short, it would be interesting to know how far the precise formal process of project selection set out by the Oswestry Hills LEADER group and the precise list of 15 selection criteria set out by the Rural Inverness and Nairn group were actually followed in practice. Reflecting on the whole LEADER process across Europe, one suspects that a good deal of 'corner cutting' was often undertaken. Certainly, the skills needed to implement rural development programmes on the ground include both an ability to decide just when the rules should be respected and when they should be bent, and the skill to bend them successfully when necessary, whilst always respecting the primacy of the underlying development mission and the imperative of financial probity.

## CASE STUDY 26: THE SUPPORT OF PROJECTS: A FRENCH EXAMPLE OF 'DEVELOPMENT VIA TRAINING'

The second case study of 'implementation' arises from the author's involvement in a transnational LEADER Observatory project to research and disseminate good practice in the use of 'training' as a tool of rural development (LEADER Observatory, 2000e). One thing that emerged during that research was a conceptualisation of 'training' by certain French development specialists very different from the more traditional British model of identifying an *individual's* training needs and then sending him or her off on a training course.

The model of *formation-développement* ('development through training') begins instead with the needs and problems of a *firm or other enterprise* and puts in place an ongoing programme of on-the-job support and training (a process called *accompagnement*). In that sense training becomes a core tool in the implementation of a development programme, much of which necessarily relates to the support of new and/or struggling small businesses and community projects.

This approach is well exemplified by ADEFPAT, the Association pour le Développement par la Formation des Pays de l'Aveyron, du Tarn, du Lot et Tarn et Garonne – four départements in the south of France that contain about 30 different agencies or structures promoting local development, including several LEADER II groups. ADEFPAT was created by a consortium of local authorities, local development associations and state agencies in the area which, collectively, perceived a need for a training support agency. It is effectively a training 'broker', wholly devoted to organising support for local development projects.

In that respect it constitutes a good example of an agency committed to the theory and practice of *formation-développement*, working to help LEADER groups and similar local development promoters deliver their objectives. But it is important to note that *formation-développement* is construed as much more than a training activity; rather it is a whole approach to teaching and learning in a developmental context.

In interview, ADEFPAT's director explained that 'the originality of *formation-développement* is that it does not start from 'training needs' as that would simply produce a classic training response; rather we start from the problems expressed by project promoters (financial problems, credibility problems, problems relating to the lack of time or to poor ways of working) and we go on from there, by interview, to translate them into "needs for greater competence" to be acquired by learning from the experience of others' (Anne Froment, interviewed in 1998).

The typical sequence of events is as follows:

1. A development agency, such as a LEADER group, proposes to ADEFPAT that it devises a training programme to support a particular enterprise or project in which it has an interest. (Individual project promoters cannot go directly to ADEFPAT to seek training assistance.) An ADEFPAT training adviser then investigates the situation to ensure that its response is genuinely 'driven by need and not by supply'.
2. If the ADEFPAT response is favourable, the development agency sets up a 'project support group' with its help, respecting the tenets of *formation-développement* and including a small number of committed people with something tangible to offer the project promoter. This support group ensures the provision of wide-ranging support and not just 'training' in a narrow sense.
3. ADEFPAT's training adviser then recruits an appropriate trainer from its network, in association with the project promoter. The trainer can be an employee of a specialist firm or training agency or a private consultant; he/ she must be skilled both in the relevant subject and as a trainer. Together, ADEFPAT's training adviser, the trainer and the project promoter work out a training plan of action, paying attention to the method, content, teaching and learning tools, timetable, evaluation criteria etc., and the training adviser drafts the contractual aspects of the arrangement. The very act of preparing this 'support based on training' is generally a sort of training as far as the project promoter is concerned.
4. A financial package, to fund the training support programme, is then put together by ADEFPAT, usually including contributions from its sponsoring agencies and various other sources which can include local LEADER funding.
5. The 'project support group' meets to approve the proposed training package and to ensure its compatibility with other types of support being offered.

6. The delivery of the training then proceeds on-site; it is the trainer who travels to the project promoter and not vice versa. Theory and practice are carefully phased so that the project promoter and his/her team can apply the theory in their day-to-day work. Typically, the training is spread over one or two years, built around the project promoter's commitments and requirements. From time to time, the project support group reviews progress and suggests any changes of emphasis in the training programme.
7. Finally, an evaluation is undertaken based on the objectives specified by the project support group at the outset. It generally has three components:

   - educational – concerning the level of competence attained and any behavioural change on the part of people engaged in the business or project;
   - economic – concerning the effect on the firm's turnover, employment etc.; and
   - developmental – concerning effects on the local 'dynamic', focusing on any spin-off effects on the local area.

The evaluation is always strongly empirical; it simply involves trying to compare the 'before' and 'after' situations. The trainer and trainees are fully involved in this exercise.

In summary, 'ADEFPAT's mission is not "to train" as such but to "engineer solutions" – to be the interface between problems and needs arising from the development process, and responses based on skills and competence. Our approach to training and learning is tailor-made "action training" for a particular project' (Anne Froment, in interview, 1998).

Reference to a specific example may be useful; this one comes from the July 1997 edition of ADEFPAT's magazine *Contact*, freely translated by the author:

> The commune of Brandonnet has had a lot of problems with its 'multi-purpose outlet' (*multiple rural*). Several managers came and went in quick succession. When the last one left, a voluntary group, 'le Bistroquet', kept the café going on Sundays, but that was all. This voluntary group and the commune council resolved that something had to change, and, advised by ADEFPAT, agreed to participate in a *formation-développement* exercise. 'We have defined with outside professional help what the real needs of the local people are, we have set up a 'support group' to pilot the project, to help the couple now running the outlet and to put them in touch with various partners so that they are better integrated into the local community', explained Dominique Ladet, deputy mayor of Brandonnet.
>
> This has led on to a 'business plan' which specifies several contributions to be made by the commune, for example relating to improvements to the building and to management assistance, and by the outlet managers themselves. 'This project has been a real spur to communication in the commune', added

Dominique Ladet 'and we are going to repeat this approach of using a "support group" in relation to other commercial projects ... it isn't possible to separate such actions from the wider process of development in the commune.'

Today Brandonnet's 'multi-purpose outlet' is flourishing to everyone's satisfaction, and bears witness to what can be achieved by intelligent collaboration between the private, public and voluntary sectors when clear objectives have been defined, lines of communication set up, and a willingness to learn has been established.

---

## *SEE ALSO...*

Case Studies 2, 17, 20, 24 and 28 also touch on aspects of implementation.

## *SELECTED FURTHER READING*

See Flanagan *et al.* (1995) and LEADER Observatory (1995b and 1998).

# 14

# Evaluation: Assessing Achievement

'Ask the fellows who cut the hay.'

Title of a book by George Ewart Evans

## DEFINITION AND IMPORTANCE

Local development, whether rural or urban in focus, is a complex, time-consuming and resource-hungry process. It follows that it is important to reflect periodically and with as much rigour as possible on whether it is achieving what has been intended, and if not, why not? It is this process of rigorous reflection upon the degree of achievement to which the word *evaluation* is generally applied.

It is not to be confused with *monitoring* – an equally necessary exercise but one much more limited in purpose, scope and complexity. Thus 'monitoring is primarily about the recording of events as distinct from understanding and interpreting their significance' (Flanagan *et al.*, 1995: 44). It is essentially a mechanistic exercise charting actions over time, for example projects started, money expended, enquiries received or meetings addressed, and asking, in effect, 'Are we on course?'

'Evaluation' is more demanding, as the following three definitions make clear:

- 'Evaluation is a systematic...analysis of performance, efficiency and impact in relation to objectives, its ultimate purpose being not to pronounce a verdict but rather to draw lessons from experience in order to adjust existing action and to modify and improve future effort.' (Casley and Kumar, 1987: 10)
- 'Evaluation...can be defined as putting a value on work undertaken over a period of time...More specifically it is about assessing whether objectives set out in a plan are actually achieved, why some activities are more successful than others, and how strategies can be improved.' Flanagan *et al.* (1995: 44)

- 'The evaluation of an action, operation, project or programme means examining it in the context in which it is applied in order to assess its effects with respect to the intended objectives.' (Campagne, 1994: 7)

Note the common word in all three definitions – 'objectives'. Evaluation is fundamentally about establishing and elucidating the degree of their attainment; as Campagne (1994) puts it, 'the evaluation process always consists of measuring deviations and analysing the reasons for such deviations' (p. 12).

To elaborate a little more on the *purpose of evaluation*, there are (after Walsh, 2000; see also Ray (2000b)) really four such purposes, each of particular interest to one or more stakeholders in the local development process:

- *accountability and value for money* – to help the promoters and funders of the programme decide whether the money has been properly spent and well-spent;
- *management* – to help the managers of the programme identify ways in which the programme's execution might be made more effective and/or efficient in the time that remains;
- *learning* – to help the policy-makers as well as various commentators and critics gain insights and understanding that might be useful elsewhere or in successor programmes – this 'learning' relating particularly to the relationships between inputs and outputs and between causes and effects; and
- *empowerment* – to enhance the skills, knowledge and commitment of those involved – this last purpose assuming, indeed requiring, that those involved in the development programme are also involved in its evaluation.

Thus the fourth of these, the 'empowerment purpose', sees evaluation as providing above all a further spur to the development dynamic. It therefore provides another example of 'the process being part of the product' in local development. Indeed, this wish that evaluation not only 'assesses' but also 'empowers' is now a tenet of local development. As Midmore (1998: 424) put it, the local development process is 'enhanced where evaluation approaches mirror the participatory, capacity building ethos of the projects themselves'.

## SOME KEY ISSUES

In the local development context then, evaluation has various purposes; it can also be carried out at various levels and at various times. As far as 'levels' are concerned, an evaluation can focus on a whole national or international programme (e.g. Ireland's entire LEADER I programme 1991–94, as evaluated by Kearney *et al.*, 1994), a specific local programme (e.g. the SPARC LEADER programme for south Pembrokeshire, already reviewed in Case Study 2), a constituent project within that programme (e.g. SPARC's 'Local Products Initiative') and a specific action carried out as part of that project (e.g. the training provided for women keen to take part in that initiative). All are valid and each may be expected to realise one or more of the 'purposes' set out above.

As for timing, evaluation can be useful

- *before* a programme or project gets underway – effectively an exercise in trying to *predict* the impact or consequences of the measures planned;
- *during* the programme or project – to identify ways of improving performance and management;
- *at the end* of the programme or project – to assess immediate impacts or outcomes[1]
- *some time afterwards* – to pick up longer-term consequences and to allow the more mature reflection that comes from 'standing back' a little.

Indeed, there is a case for evaluation being undertaken more or less continuously during the life of a programme or project, as a part of the normal process of management, though ongoing evaluation of this kind can be expensive and distracting.

The evaluation of a local development programme is rarely easy. Consider the challenge of evaluating an integrated rural development programme that has been running for three or four years, say, in an area with about 50,000 people spread across perhaps 1000 sq. km – a typical LEADER area. A first problem is that for such an area there will almost certainly be inadequate statistics (on employment, unemployment, incomes, social deprivation etc.) at a sufficiently disaggregated level for either the 'baseline year' – when the programme started – or for the present day (three or four years after it started). Couple that with a second problem, that across such a wide area any effects of the development programme will almost certainly have been overlain, if not swamped, by the consequences of other unrelated events, such as an upturn or downturn of the national economy, and it becomes virtually impossible to say 'subtract the baseline situation from that pertaining today and you have the net effect of the programme'.

With that avenue of enquiry blocked, the researcher is forced to focus instead on the specific actions of the programme, trying to trace them through to their consequences on the ground. But again there is the problem that other factors will certainly have 'muddied the waters'. The most striking example of this is the 'jobs created myth' – the temptation to say, in effect, that 'There are 150 people working in the 12 factory units that have been built with funds made available by this programme, therefore we have created 150 jobs.' Almost certainly, such a statement will be false; it ignores both the 'deadweight problem' (meaning the extent to which the 150 jobs would have come about anyway) and the 'displacement problem' (meaning the extent to which the 150 jobs were already in existence in other, even if less suitable, premises elsewhere in the area).

More fundamentally, local development programmes are complex and multi-dimensional processes; they generally comprise multiple actions, pursued by multiple actors in pursuit of multiple goals. Just 'what affected what' is difficult to entangle – a problem exacerbated by the often intangible nature of some of the outcomes being pursued in local development – for example a 'culture of enterprise' or a 'greater sense of local identity'. Moreover, local development is by its very nature a long-term process, whose full effects may be expected to emerge only after many years. What *these* problems – the problems arising from programme complexity, inherently intangible outcomes and delayed consequences – tend to produce is one of two unfortunate results: either the valuable outcomes of the programme get to be overlooked by an 'over-technocratic' evaluator concerned only to put ticks in boxes; or else they are exaggerated by a rather complacent programme manager, conscious perhaps of the lack of more tangible and visible outcomes on the ground.

Given all these difficulties it might be tempting to conclude that a genuinely serious and useful evaluation of a local development programme is impossible – but obviously that is not an option. An assessment is still needed of whether this or that programme or project is proceeding well and is proving worthwhile. What the difficulties listed do ensure is that, where local development is concerned, two sorts of 'evaluation' are really of very little value. The first is a simple comparison of 'before-and after snapshots', whether they be of the state of the whole geographical area or of certain key indicators of the 'jobs created, business started, trainees trained' variety. The second are formal analytical approaches of the 'cost-benefit analysis' type that are often quite usefully applied to defined sectoral programmes in agriculture, for example, or to one-off discrete local projects such as the construction of a bypass around a certain town. Local development demands the use of other tools.

## THE TOOLKIT

With such approaches proving of limited value, what has happened in recent years is that most evaluators of multi-dimensional, multi-purpose, multi-project local development programmes have tended, in effect, to endorse Campagne's contention that the task is really to focus on three questions – What were, or are, the programme's objectives? How far were, or are, they being realised? And Why/why not? (Campagne, 1994: 18). The importance of the third of these questions (why/why not?) is that it clearly forces attention to be focused on the alleged 'chain of causation' between the act and the consequence, and it is an elucidation of that chain that really makes the exercise worthwhile.

It follows from all of this that an appropriate methodology for the evaluation of local development programmes or projects is likely to be

- more qualitative than quantitative, while not neglecting the measurement of whatever can usefully be measured; and
- strongly participative, involving the stakeholders both to achieve some learning and developmental spin-off (it is almost inconceivable that, genuinely involved in the evaluation, the stakeholders will not themselves be 'developed' in some way) and because it is those who are closely involved, if carefully questioned in a non-threatening way, who are most likely to throw light on the crucial 'why/why not?' question.

Such an approach sits comfortably with Campagne's assertion that the evaluator's task is to go back to the objectives of the project or programme to see how far they have been met and then, using a qualitative methodology, to try to establish how far that success or lack of it is attributable to measures taken in the programme. Equally, it sits comfortably with Midmore's conclusion that 'indicators of this success (meaning success in the complex process of local development) can be had only from a combination of measurements of such tractable indicators as exist together with an intuitive but equally careful analysis of non-quantifiable information' (Midmore, 1998: 419).

This conclusion – that in most cases an appropriate methodology for the evaluation of local development programmes and projects might usefully focus upon 'objectives; their attainment; reasons for non- or partial attainment' – requires us to consider two further issues before an attempt is made to spell out the essence of a basic model. The first concerns qualitative analysis; the second the use of 'indicators'.

### *QUALITATIVE ANALYSIS*

Qualitative analysis (that relating to non-numerical information) is increasingly applied in such circumstances, not just because the necessary numerical data are simply unavailable or of dubious validity, but because deeper insights can often be achieved if the careful elaboration of the direct experience of those involved becomes central to the analysis. In practice, this means, quite simply, dealing with words and lots of them. This is particularly the case when the programmes and projects in question have been explicitly devised to achieve a qualitative change in attitudes and culture as much as the creation of jobs and an increase in incomes.

Lest this be construed as the sacrifice of objectivity in favour of the mere culling of opinion, it needs to be said that social science today is loath to accept that 'quantitative means objective' while 'qualitative means subjective'. So-called 'quantitative objectivity' raises serious questions about the researcher's own subjectivity in defining the scope of the analysis, interpreting the concepts, selecting the method, assembling and rejecting data, interpreting the analysis etc; and qualitative analysis may well be best for elucidating the various 'objective realities' constructed by the human beings who are the essence of the whole business of development.

Either way – quantitative or qualitative – the researcher is, in practice, faced with the challenge of trying to assemble as much relevant evidence as possible according to a well-constructed analytical framework, and then interpreting it and expressing it with as much care, transparency and integrity as possible. But whatever approach is taken, subjectivity is bound to be present.

Assembling source material for the qualitative analysis of local development generally involves three exercises:

First, documentary analysis. Relevant documents might include, for example, business plans and their updates, internal progress reports, project funding applications, background papers for committee, diaries, time-sheets, the minutes of meetings, newsletters, press cuttings and correspondence. It is surprising how much can be gleaned about a programme and about why it has done well or not so well simply by sitting down for a week and reading and noting within a structured analytical framework.

Second, semi-structured interviews. These allow the interviewee to elaborate freely around preselected themes. Relevant interviewees will almost certainly need to be drawn from among each of the programme or project principal stakeholders – its funders and promoters, its partners and managers, its beneficiaries on the ground and what might be termed its close observers – those with an interest but no direct involvement. Some such interviews can be of groups collectively rather than of individuals alone.

Third, direct participant observation. Often this involves checking whether people actually behave as they claim. For example, is the programme management committee *genuinely* open and inclusive or is it dominated by a clique or by the salaried officers? Do the programme's field officers *genuinely* reach out to excluded people or do they, in practice, cultivate their existing contacts? One way of finding out in each case is to watch and to listen.

Each of these techniques is likely to produce a great many words to reflect upon subsequently. This is not the place to expand on the treatment of textual material; suffice to say that, in essence, the task is to structure all the material around themes and questions that are largely predetermined by the purpose of the exercise but which may also emerge as one reads and re-reads. In this process, the elusive goal of 'objectivity' can often be enhanced if *two* researchers go independently through the textual material and then see if they have arrived at similar conclusions.

### THE USE OF INDICATORS IN THE APPRAISAL OF 'DEVELOPMENT'

Notwithstanding the value – indeed often the inevitability – of employing a largely qualitative methodology in evaluations of local development programmes,

**TABLE 14.1** ***An attempt to translate intangible objectives into some tangible indicators***

| Some objectives espoused in the local development programme | Examples of possibly relevant indicators |
|---|---|
| 1. To achieve some 'capacity building' | • the number of people directly involved in community activities who were hitherto not involved |
| 2. To foster innovative projects | • the proportion of funded projects that have no earlier parallel in the area |
| 3. To pursue sustainable development | • any evidence, for each project supported, of *adverse* effects upon the area's endowment of natural, manufactured, human or social capital<br>• likewise, any evidence of *positive* effects thereon |
| 4. To create a culture of enterprise | • the proportion of project proposals, whether subsequently funded or not, coming from people with no previous business experience |
| 5. To foster truly integrated development | • the number of funded projects linked in some way to other funded projects |
| 6. To mobilise the community to play a more active part in the local area's development | • the number of 'ordinary people' (however defined) genuinely involved in the development group's own decision-making |

there is also a strong case for trying to express in some sort of quantitative form the more intangible phenomena with which one is confronted. Often this is most useful when faced with defining the degree of attainment of the programme's objectives. Table 14.1 illustrates this point.

The point is not that seeking to re-express goals and objectives in 'quasi-quantitative' indicators and then that seeking out data in relation to them will somehow 'provide the answers'. Rather, the very process of carrying out that exercise and of using the results, however imperfect, as a spur to informed discussion is likely to clarify whether, how far and why those goals are actually being attained. For example, 'One of your key goals has been to develop a local culture of enterprise, but it seems that about three quarters of all the projects you have funded have come from established businesses. What do you think about that?'

To summarise: the best way to evaluate most local development programmes is to focus on their objectives, assess the degree of their attainment and then try to seek out and understand the complex processes of cause and effect that underlie that attainment or lack of it. In this, the evaluator must keep sight of the multi-purpose, multi-action and multi-actor nature of local development, quantify what can be quantified and delve deeply using mainly qualitative methods into the how? why? and why not? Going on from that, the precise choice of the evaluation method will depend on many factors including the real purpose of the evaluation, the complexity and scale of the programme, the availability of data, the likely co-operation of the various stakeholders, the evaluator's resources of time and money etc.

But some guiding principles are:

- Keep the focus on outcomes, intended and unintended, and on the processes that led to them; quantify what you can but do not ignore what you cannot; involve the various stakeholders as 'subjects' in the evaluation and not just as 'objects'.
- With careful reference to documentation and by means of initial exploratory interviews, start with the originally intended goals, objectives and intended outcomes, tangible or intangible, explicit or in some ways implicit.
- Try to refine those goals, objectives and intended outcomes into more tangible indicators. Then as far as possible, using eclectic sources, make a first attempt at defining how far they have been or are being attained, or at least whether 'things are moving in that direction'.
- Use that evidence as a focus for a retrospective exploration of 'what happened' – of how and why goal attainment was achieved or missed. It is at this stage that semi-structured interviews with a wide range of actors, documentary analysis and participant observation will be invaluable.
- Synthesising and critically appraising the possibly divergent evidence emerging, attempt an exposition of the links between the inputs and outputs and the causes and effects that are at the heart of a development programme.
- Test out that draft exposition by revisiting your sources and seeking their reactions.

We will now consider how far those general principles are apparent in two very different evaluation exercises.

## CASE STUDY 27: AN EVALUATION OF THE MARCHES LEADER II PROGRAMMES

One example of the many evaluations of Britain's LEADER programmes undertaken in recent years is that of the Marches group of four LEADER II programmes, namely those for Oswestry Hills, Clun Valley, Teme Valley and

Herefordshire Hills. CSR Partnership, working jointly with researchers from the University of Birmingham, were commissioned to do a 'mid-term review' of the four programmes which, collectively, served areas totalling some 2400 sq. km of the most rural parts of Shropshire, Herefordshire and Worcestershire, and which contained about 83,000 people. By mid-1999, the time of the study, the four programmes had been running for a little over half of their allotted six-year time-span.

The consultants were required to do two things (CSR Partnership, 2000: 2):

- to evaluate the success and impact of the four programmes, including an assessment of the 'seven distinctive features/processes' of LEADER; and
- to provide recommendations for a 'succession strategy' in the light of an anticipated competitive call for tenders for a follow-on programme – 'LEADER Plus' – which would start in 2001/2.

In short, the four local action groups which managed the programmes, together with the local, regional and national agencies that oversaw them, wanted to learn how well the programmes were doing in the context of the initial expectations, how their performance might be fine-tuned in the time remaining and what ideas and lessons might be gleaned to help frame decisions relating to the next phase. In this brief review we will focus on the consultants' attempt to answer the first two questions: 'How well were the programmes doing?' and 'What modifications might usefully be made?'

The consultants were faced with the task of appraising a programme comprising a total of 161 funded projects spread across the four areas. Examples of some of those projects to be subjected to detailed investigation as the work proceeded were: a small-business grant scheme in south Shropshire; an IT training programme delivered in a number of community resource centres; a village-based family centre providing childcare and other support; an initiative to launch a range of organic dairy products under a single brand name; a project to develop six trails for walkers, cyclists and riders; and a bracken composting scheme with both environmental and commercial objectives.

Thus the projects were certainly diverse, and so were the 'outputs' that project managers were committed to deliver in return for LEADER funding – training courses run, childcare places provided, tourists attracted, hectares of land improved, etc. The problems that the evaluators faced included the difficulty of identifying tangible impacts and outcomes, over and above such crude counts of 'outputs' (see note at the end of the chapter for a definition of these terms) after just a couple of years' operation in the case of most of the projects. Not surprisingly, they opted for a methodology that began by trying to quantify outputs as far as possible, and then went on to seek qualitative evidence of the broader outcomes and of the factors and processes that lay behind them.

They undertook six exercises:

- an analysis of documentation in the LEADER groups' files;
- attendance at a number of management meetings of those groups;
- a postal/telephone survey of all 161 projects, with a 71 per cent response rate;
- face-to-face interviews with 21 of the project promoters or managers;
- face-to-face interviews with 12 'key stakeholders', namely officers and members of the local action groups and representatives of partner agencies; and
- four concluding workshops, each with about 20 widely drawn participants, designed to serve as sounding boards for the consultants' provisional conclusions and as a way of speculating on longer-term options.

In fact, 'capturing the success of projects via output measures proved difficult' (CSR Partnership, 2000: 4). It was certainly possible, by scrutinising the files and perusing the questionnaires returned by the majority of the individual projects, to state that in the Clun Valley, for example, 14 tourism promotion initiatives had so far been undertaken (compared with a promised target of 19), 12 enterprises 'assisted' (compared with a target of 224) and three 'permanent jobs' created (compared with a target of ten). But we are left wondering as the local project promoters clearly did – and the consultants also we suspect – about what such highly dubious statistics actually mean. In fact, the postal/telephone survey of the individual project promoters seems to have revealed more about their real feelings – positive and negative – about the LEADER programme as it had benefited, and in some respects frustrated, their work than it did about the project's contribution to local development.

More of real value came from the face-to-face interviews with the 21 project promoters and with the dozen stakeholders. It was in these interviews that the consultants got most of what they termed the 'softer/anecdotal evidence of success' (p. 70) and a real feel for some of the community and economic development that was being achieved, as well as for the day-to-day challenges of resolving problems of cashflow, co-finance and bureaucracy.

As for assessing the four programmes against the 'seven distinctive features/processes of LEADER' (a list broadly similar to that set out in this book's Case Study 1), again the approach was essentially to try first to quantify the quantifiable and then to delve further in the various interviews. For example, the validity, in practice, of the 'bottom-up approach' was first checked out in the questionnaire survey, which revealed that a high proportion of the projects had either stemmed from ideas first mooted by 'the community' or else had been worked up by sponsoring agencies with the substantial involvement of the local community. It was then explored further in the interviews which revealed the crucial role of the LEADER programme's project officers in soliciting and then supporting 'grass roots' ideas. Without such a person a much more top-down approach would have been inevitable.

Similarly, with 'innovation', another key plank of the LEADER philosophy, the survey of project mangers established that in the case of 62 per cent of the projects (for example the credit union project and another promoting distance learning for the voluntary sector) nothing similar existed locally when they began – in other words they were 'innovative'. Then the in-depth interviews were able to add more detail and a critique. For example, the interviews produced evidence of innovative commercial links developing across the adjacent Welsh border. They also produced the intriguing suggestion that, on occasions, project funding applications had been refused largely because no-one had ever tried the suggested approach before – a clear negation of the objective of fostering innovation.

To conclude, this mid-term evaluation of four neighbouring LEADER programmes exemplified many of the challenges of evaluating multi-purpose, multi-project, multi-agency, area-wide local development programmes. In the end, clear-cut statements of success or failure, or composite measures of impact remained elusive. Rather, by the judicious use of complementary research techniques, quantifying where possible, but never being duped when alluring tables of figures emerged, and always probing and synthesising the perspectives of the different stakeholders, a valid and valuable picture emerged. In this case the evaluators were able to draw reasonably firm conclusions both about the initial effects of the programmes and about the strengths and deficiencies of management and delivery which were subsequently useful in the final years of the programmes and in the consideration of future options.

---

## CASE STUDY 28: AN EVALUATION OF 'RURAL ACTION' ASSESSING THE COMMUNITY DEVELOPMENT SPIN-OFF OF ENVIRONMENTAL CONSERVATION

---

Our second case study relates to a programme to support community-led local environmental action across rural England. Running from 1992 to 1999, Rural Action was funded, significantly, by a consortium of three national agencies, each with a rather different 'take' on the countryside and little prior experience of working together – the Countryside Commission, the Rural Development Commission (the two of them later being fused into the Countryside Agency) and English Nature. In a formal 'statement of joint commitment' (reproduced in Taylor, 1998) the chairmen of the three agencies set out their conception of the programme – community-led action to conserve the environment and encourage its sustainable use, and environmental action as a means of strengthening rural communities.

That duality of purpose – local community action to support the rural environment and local environmental action to support the rural community – underlay another key characteristic of the programme. There were to be 40 'county support networks', each assembling a range of expertise from appropriate local authorities, voluntary bodies and statutory agencies (e.g. the county-level Rural Community Councils and Wildlife Trusts) and charged with helping action groups at the very local level to plan and carry out their conservation work to maximum effect. 'Network development grants' were made available to help these county networks to gel and to do their promotional, training and support work, as well as 'project grants' to part-fund the local action groups' work on the ground.

By the time that Warburton (1998b) was commissioned to undertake an interim evaluation of the programme, some 3770 individual projects had been funded – an average of almost 100 per county. This was at a cost to the exchequer of only £3.2 million (less than £900 per project on average), plus local 'matched funding' of rather more than that amount. Importantly, this 'matched funding' could, and did, involve a good deal of voluntary labour, costed in at a shadow price of £46 per day. The projects were diverse – local surveys of wildlife, energy audits of community buildings, practical conservation tasks, recycling and composting initiatives, the creation of village trails, etc. So, too, were the local groups undertaking them – parish councils, village hall committees, residents' groups, conservation groups, local history and arts groups etc., as well as *ad hoc* groups coming together for the task in hand. Significantly, most could be described as essentially 'non-environmental' with the implication that a broader constituency of activists was being attracted into 'conservation work' of one sort or another.

In the present context, our interest in Warburton's evaluation lies in its brave attempt to use largely secondary sources (there being no time for original surveys) to establish how much community benefit had spun off from this 'environmental' initiative in its first five years. More specifically, she was given just three months and a very limited budget mainly to assess the *community* effects of the programme rather than the more tangible environmental improvements, which were what the local groups were aiming to achieve. Her analyses were largely qualitative but they did involve creating a number of 'indicators' and making an attempt to measure the success of their attainment. Her sources comprised largely a mass of written material assembled from the county networks, various partial evaluations already undertaken, often at the county level, and over 100 telephone interviews with activists in the county networks or in local projects.

Her central concern was with 'capacity building'. How far had Rural Action actually built up the capacity of local people and groups to undertake useful 'community work' in its broadest sense, over and above the particular projects to which the grants related? With that in mind she focused on four indicators:

- *the formation of new groups* Here the evidence suggested that, depending on the sources used, between 20 per cent and 60 per cent of the local groups had come into existence specifically to undertake a project with Rural Action funding – the implication being that the programme had therefore helped to create some valuable social capital.
- *participation within the groups* Here the evidence was more sketchy but suggested that some 70,000 individuals had been involved in some way in devising or delivering the 3770 projects. More tangibly, some 95,000 person-days of voluntary effort had been costed into project budgets as part of their 'matched funding'.
- *the amount of learning achieved* With regard to this indicator, Warburton was compelled to use 'training purchased' as a proxy measure and was able to establish that about half of all the local groups had spent some of their budget on training. Often this related to research, planning or project management skills which would be transferable to subsequent community ventures.
- *the extent to which the groups went on to undertake further projects* Here it was established that between 30 per cent and 60 per cent of local groups went on to undertake further projects once the Rural Action money had been spent. (One example cited involved a pond restoration project that led to a parish map of valued local features and then to a full parish appraisal.)

All of this points to a measure of success in relation to a key objective of the Rural Action programme – 'environmental action as a means of strengthening communities'. But clearly, Warburton's necessarily 'quick and dirty' evaluation cried out for subsequent detailed local research to test out and flesh out the tentative conclusions that she had reached. Moreover if, as previously suggested, the essence of evaluation in a rural development context really does amount to answering three questions – Objectives achieved? If so how? If not, why not? – then it is clear that the job was only partly done. She was unable to 'dig deep' into the causes of Rural Action programme's partial success in fostering local environmental action as a tool of community development. But she had revealed some measure of success in that respect and done enough to refute the frequent assertion that the so-called 'soft outcomes' of rural development defy careful evaluation.

---

## NOTE

1. Four terms commonly used in the evalution of local development programmes warrant definition: *Outputs* are the tangible results of actions – such as the number of training places provided by a local LEADER Programme. *Outcomes*

are the less tangible consequences of achieving those outputs – such as a more skilled labour force. (Funding agencies often want the latter but feel forced to count the former.) *Effectiveness* is a measure of the attainment of objectives. *Efficiency* refers to how well resources are used in that respect.

### SEE ALSO...

Of the other case studies, only number 4 relates explicitly to an evaluation, in that case of a range of projects and programmes appraised against sustainability indicators. In addition, see numbers 17, 20 and 23 for three case studies which were largely the product of evaluations by the author but which say little of evaluation methodology.

### SELECTED FURTHER READING

Midmore (1998) provides an excellent overview of issues surrounding the evaluation of local development. Flanagan *et al.* (1995) and LEADER Observatory (1995b) provide practical guidance. Kearney *et al.* (1994) provide an evaluation of a rural development programme.

# 15

# Conclusion: More Research Needed

> This day has ended.
> It is closing upon us even as the water-lily upon its own tomorrow.
>
> Kahlil Gibran

It is not intended to repeat here the arguments advanced in the earlier chapters for 'local rural development' or to reassert the significance of a number of key concepts in efforts to promote it. Nor will any attempt be made to pull those ideas together into an all-embracing theory of such development. Rather, we will make a few suggestions for further research which stem from the earlier discussion, and in doing so necessarily point up some still unresolved issues which practitioners of local development might usefully bear in mind as they proceed. These research suggestions are grouped together by chapter, using the 'strap-lines' incorporated into each chapter title as an alternative to the one-word descriptors of the 13 concepts or themes. No further reference to the literature will be made, the reader being invited to look back at the preceding text to link up with some of the relevant research already undertaken.

## MAKING IT LOCAL

As was suggested in Chapter 1, while the case for promoting rural development within defined local areas is central to our argument, there is no hard and fast rule for deciding 'How local is "local"?', much less for defining the specific local area(s) to focus upon. It would therefore be useful to examine a large sample of recent local development programmes with the aim of establishing just how relevant to their performance were the size of the areas they served (both population size and areal extent), their cohesion and their degree of congruence with pre-existing administrative boundaries. What, in practice, was the validity and pertinence of each of those parameters in the framing and delivery of local development?

In addition, in what ways and to what extent is a local focus insufficient or even in some respects inappropriate? Just as 'no man is an island' (John Donne's celebrated comment on society, reproduced at the start of Chapter 6) so 'no local area is an island' in an economic, social, cultural or political sense. Interaction with the world beyond the locality is inevitable and in some respects essential, and 'self-containment' as an objective is ultimately illusory. While that reality is generally recognised at least tacitly, the nature of the extra-local links needed to complement a locally focused development programme could be usefully explored and clarified. Relevant in this respect would be links with external markets, institutions and networks and various kinds of transnational collaboration.

## RESPECTING THE LONG TERM

Chapter 2 argued that local development should respect the need to conserve and, if possible, enhance the 'four capitals' – environmental, manufactured, social and human. Students of this 'sustainability imperative' have tended to focus rather more on environmental capital than on the other three, and here we suggest a need for more work on social and human capital as casualties of inappropriate local development or as welcome products of appropriate development. Taking the experience of the last 20 years, for example, what evidence is there of the erosion, or alternatively the enhancement, by local development programmes of, say, the extended families, informal care networks and cultural activities allegedly once commonplace in England's villages? And what were the consequences of that erosion or enhancement in a developmental sense?

A second 'sustainability issue' raised in Chapter 2 concerns 'transfrontier responsibility' – and here there is an obvious connection with the point already made about the non-local dimension of local development. In particular, does championing local distinctiveness as a strategy of local development necessarily imply greater long-distance and resource-demanding interaction with distant markets – including the attraction of car- or air-borne tourists? Thus, does what may be loosely termed the LEADER model of local development – as outlined in the pages of this book – imply some inevitable 'export' of unsustainable behaviour and, if so, what might be done about it?

## BREAKING THE MOULD

We argued in Chapter 3 that the adoption of innovation is an essential component of any development programme, with 'more of the same' not constituting 'development' in any fundamental sense. The argument is partly that 'it is the grit in the oyster that makes the pearl' and that the adoption of innovation typically provokes some measure of social and cultural change – a change in attitudes, values and ways of life – which *is* a component of development. But just how this change occurs and how it confers local benefits warrants closer study. In effect, this is a call for the careful study of the various and varied consequences of innovation in a local development context. A start could perhaps be made by considering the increased use of information and communication technology in service delivery in rural areas or, perhaps, the increased tendency for public bodies to deliver their services via contractors rather than directly. What are the wider social and cultural consequences of such innovations, as well as the more narrow economic ones? And do any such consequences further stimulate, or constrain, development?

A second suggested piece of innovation-related research would be more straightforward – it would involve 'simply' trying to describe and explain the diffusion over time of certain development-inducing innovations. Examples might be the spread of attempts by local communities to produce village design statements, or of post offices located in pubs, or of ring-and-ride transport schemes, in each case concentrating on the experience of a given region over a given period of time. There is an enormous literature on the adoption (or non-adoption) of innovation by farmers, but very little relating to other rural actors. Chapter 3 gave some pointers on why and how such research might be done.

## BUILDING ON WHAT'S THERE

If a development agency is genuinely committed to a local development approach then it must surely be involved in a systematic appraisal of the economic potential of the full gamut of local resources, physical and human. But how best to do this? Research could usefully focus on alternative methodologies for identifying the scope for adding value locally, with Chapters 4 and 11 both suggesting ways of doing this which require further refinement and empirical testing.

A related issue is the scope for 'plugging leaks' in the local economy by inducing local businesses and other organisations, as well as households, to purchase more from local suppliers and thereby to increase the local multiplier. Chapter 2 broached various practical ways of doing this and referred to some action-research being undertaken by the New Economics Foundation. But two more fundamental questions also warrant attention: first, is 'plugging leaks' a strategy that only a few localities can successfully pursue since champions of local development tend simultaneously to advocate the vigorous export of goods and services from the locality – exports that must necessarily be imports elsewhere? and second, does the increased self-containment of local economies arising from plugging leaks, bring lower productivity in its train by reducing the incentive for local areas to specialise in those economic activities for which they have a comparative advantage? Putting such concerns alongside those already raised in relation to 'transfrontier responsibility' does not negate the strong and multifaceted case for *local* development, but it certainly suggests some caveats that need to be clarified.

## BACKING THE RISK-TAKER

Moving on to consider the fostering of entrepreneurship, a distinction was drawn in Chapter 5 between commercial and community entrepreneurs, though in each case the challenge is to increase their number and effectiveness in given localities. In the case of commercial entrepreneurs, there has been insufficient research into the relative merits of different ways of generating a more entrepreneurial culture in areas where such a culture is deficient, and of seeking out and supporting those individuals who might be prepared to 'have a go' at setting up new businesses or expanding existing ones. Development agencies try different ways of doing these two tasks and some comparative research into their effectiveness would be helpful.

As for 'community entrepreneurs', the evidence suggests that there are often too few of them in relation to the opportunities for involvement now being given by central and local government to local communities – or, perhaps, *imposed* by government on local communities. Basically this is a call for research into capacity building in village communities, but more specifically it is a suggestion that more be learned about the characteristics, motivations and behaviour of local community leaders and of those who might be persuaded to join or replace them.

## FOSTERING A SENSE OF BELONGING

With regard to 'community', Chapter 6 suggested a need to better identify its existence and quality as manifested in local areas and to relate that existence and quality to possible correlates or even causal factors. This might mean looking at a sample of villages or parishes, developing, however tentatively, various indicators of the 'community' to be found there and positing relationships between those indicators and a range of possible 'explanatory factors'. The latter might include the localities' social mix, the pace there of recent change, the existence or absence of such meeting-places as pubs and village halls, the recent involvement of community development workers and any recent empowerment initiatives. To this end one might suggest both a large-scale quantitative study involving scores if not hundreds of localities and more in-depth, qualitative case study work.

It would also be useful to know more of the economic value of 'community' in the local context. This might involve, for example, trying to estimate the additional costs that would necessarily fall on the state if local volunteers were not involved in good neighbour initiatives, neighbourhood watch schemes and community transport provision, say, or in managing village halls and community enterprises. Such an exercise might inform the calculation of how much resource should be put by the state into local community development and capacity building initiatives.

## BRINGING ON BOARD

Chapter 7 was devoted to 'social inclusion' and suggested that excluded people might be better construed as an under-utilised resource than as a problem to be alleviated – making them, in effect, 'subjects' in a local development programme rather than merely 'objects' of concern. The resources that, for example, the long-term unemployed, ethnic minorities, lone parents, women at home and elderly people living alone might each bring to bear needs systematic research; those resources could include particular local knowledge, certain life skills, and a capacity for certain kinds of community care and community enterprise. Related to this is a need to know more about the machinery that would be needed at the local level to exploit or 'valorise' such resources.

There is also a need to know more about how best to involve 'hard to reach' groups and 'non-joiners' in exercises of consultation and participation. Such involvement would have a dual objective: ensuring that their contribution is heard alongside that of their more articulate neighbours, and indirectly developing their confidence and abilities. Attempts to achieve such involvement generally indicate a need to meet the 'hard to reach' on their own ground and to respect their own agendas, and often this means intensive work by skilled and trusted facilitators linked, say, to childcare and women-returner initiatives, and to youth projects. More 'action-research' in this respect would be useful.

## BRINGING WITHIN REACH

Moving on to the 'accessibility problem' as explored in Chapter 8, a key need is to know more about alternative community-based management and monitoring systems at the local area level. (Case Studies 15 and 16 both explored the difficulty of ensuring a holistic approach to 'accessibility management' at the sub-county level.) This is a classic arena for the reconciliation of top-down and bottom-up approaches to planning and management – each is essential – and comparative research in this area would pay dividends. But it has to be clear that 'accessibility planning and management' means much more than just the better co-ordination of public transport.

Chapter 8 also suggested that information and communication technology (ICT), as it is increasingly deployed in the delivery of rural services, is a double-edged sword. Several examples are cited there of ICT helping service providers to sustain small and very local outlets and increasing the development potential of remote rural areas. But the same technology also risks eroding the cohesion of local communities and increasing social isolation in various ways. 'ICT at the service of rural communities and local development' is, therefore, another theme warranting research.

## WORKING IN HARNESS

As for local partnership working, the core question is what do partnerships bring to local development that agencies and local groups working alone do not and could not bring? A related issue concerns how to increase partnership 'value-added' in the local development process. Those seemingly simple questions conceal a mass of considerations, some of them outlined in Chapter 9, based on the recent 'PRIDE' research project and other work. Further empirical research, teasing out the sources of partnership value-added in relation to real-world experience, and thereby identifying and disseminating elements of good practice, would be useful.

Most research on local partnerships has focused on those which have come about essentially as a tactical response to the existence of new funding opportunities or as a reaction to official dissatisfaction with the effectiveness of single agencies working alone. Such partnerships, deliberately brought about by state or local government influence, may well comprise the majority. But there is a need for more research into the formation, operation and effectiveness of 'spontaneous bottom-up partnerships', such as many community development trusts, which typically involve a number of local people coming together to discuss and to seek to resolve local problems without any expectation of external funding, at least initially.

## EMBRACING THE PEOPLE

In outlining a 'consultation and participation toolkit', Chapter 10 could do little more than touch on the wide range of techniques now employed in practice. In some respects this is a well-researched subject but there has been little systematic consideration of the following question: 'Which techniques are most effective in delivering "capacity building" as a by-product of consultation and/or participation, and how and why?' Research thereon would involve looking carefully at a range of local consultation/participation exercises undertaken in the recent past, say three to six years ago, establishing the subsequent activities of many of the consultees/participants, and trying to relate those activities to the experiences that they had had in the earlier period.

We move on now to set out some research suggestions flowing from the four 'practice' chapters of the book.

## RESEARCHING THE BASELINE

Chapter 11 and its two related case studies outlined a number of techniques of 'area-diagnosis' which incorporate various combinations of technical analysis and popular involvement. In relation to the actual employment of such techniques, and others, it would be valuable to research just how they have been applied in practice and, more particularly, to establish how and how far they subsequently influenced the local development strategy and action on the ground. For example, considering the Countryside Agency's 'Market Towns Health Check', how successfully have local communities executed what, on paper, is clearly a demanding exercise of information assembly and analysis? And what links can be traced between the lessons that were learned about the town by carrying out the 'health check' and subsequent social, economic or environmental initiatives and developments?

A more specific piece of research would relate to one particular technique, the culling of 'local testimony'. Again, what are the best ways of gathering from 'ordinary' local people useful information and opinion on development needs and opportunities, and how can that testimony be best integrated with other information and then translated into proposed action?

## ORCHESTRATING ACTION

Moving on to the formulation of strategic plans for the development of defined local areas (the subject of Chapter 12), a key need is to explore alternative ways of marrying the 'top-down' and the 'bottom-up', after a rather long period in which an underlying assumption has tended to prevail that in local development the former is 'bad' and the latter 'good'. The need for a better synthesis of the two approaches to formulating strategies for local development is already becoming apparent in relation to the preparation by England's local authorities of holistic, area-wide 'community strategies'. These must build upon, but not just passively collate, needs and priorities being expressed at the parish or village/small-town level. Again, research might usefully explore the range of practice which is developing.

Case Study 24 focused on development planning at the very local level; this has come to the fore with the government's 'parish plans' initiative. One issue concerns how such plans actually emerge, or might best emerge, once an appraisal or other diagnostic study has been undertaken. In the author's experience they tend to 'fall from the sky', in the sense that local communities seem often to leap from appraisal to plan in a way that must leave doubt regarding the justification of the measures proposed. Whether such creative leaps can be usefully formalised in some way seems to be an important question if local communities are to be taken seriously as 'bottom-up' players in the local development process.

## MAKING THINGS HAPPEN

As suggested in Chapter 13, as far as the implementation of local development strategies is concerned, there seems to be a gradual and welcome shift away from 'project-ism' (the assumption that implementation simply means funding appropriate projects) to 'influencing', meaning the persuasion of other agencies to modify their actions so as to accord better with the local development strategy. Just how, and how successfully, such 'influencing' works in practice would repay careful study, as would the suspicion that inequity can creep in at this point as a consequence of the better prepared and more articulate communities having a disproportionate ability to influence the deployment of resources.

Returning to the selection of projects as a tool of implementation, it would be interesting to know just how far, and how, checklists of criteria (such as those set out in Case Study 25) are *actually* used by, for example, LEADER groups to guide the choices they have to make; and can good practice in this field be crystallised, in relation not just to the use of selection criteria but also to the ability of such scrutiny to help develop and hone project proposals, as distinct from simply assessing them? There are obvious links here with the discussion of sustainability, quality of life capital and related matters in Chapter 2.

## ASSESSING ACHIEVEMENT

In effect the previous paragraph was a call for better *ex-ante* evaluation techniques – those techniques which seek to assess the likely impacts of proposed projects and programmes before they are launched. There is also a need for better *ex-post* evaluative tools – the word 'better' implying two things: first, an improved ability to capture the less tangible 'outcomes' of development initiatives as well as their more tangible but often deceptive 'outputs' (see the distinction suggested as a footnote in Chapter 14); and second, an improved ability to achieve 'capacity building' through the appropriate involvement of local actors in the evaluation exercise.

Just how to achieve those *desiderata* of the evaluation of local development while simultaneously helping hard-nosed funding agencies to judge whether they have received value for money, is a question that could keep researchers busy for some time to come.

At the end of this swift *tour d'horizon* of research priorities, we return to the elusive, indeed as far as this book is concerned, side-stepped goal of a comprehensive and operational theory of local rural development – certainly a socially worthwhile goal if we recall Bertrand Russell's celebrated maxim that 'There is nothing more practical than good theory.' How does the whole edifice of local rural development hang together? What causes what and how? The chapters of this book have given some hints in that respect but little more.

Useful progress towards a comprehensive, practical and validated theory of local rural development might, however, be achieved by means of an empirically based programme of work devoted to teasing out the nature and force of the links between

- real-world efforts to pursue, within local development programmes, the nine goals implied by the titles of Chapters 2–10; and
- genuine 'development on the ground', interpreting the term 'development' as it was defined early in Chapter 1.

Such work might focus on the experience of a large and varied sample of local rural development programmes undertaken in Britain and elsewhere in Europe in recent years. As and when accomplished, its findings should be disseminated in plain language and to all who might need to know.

# References


ADEFPAT (1996) *Nouvelles pratiques de l'Insertion-développement en milieu rural.* Paris: Grep Editions.

Area Development Management (1995) *Integrated Local Development Handbook.* Dublin: Area Development Management.

Arnstein, S. (1969) 'A ladder of citizen participation in the USA', *Journal of the American Institute of Planners,* 35: 216–24.

Atlantic Consultants (1999) *South West LEADER Forward Strategies – Final Report.* Truro: Atlantic Consultants.

Beatty, C. and Fothergill, S. (1999) 'Labour market detachment in rural England', *Rural Research Report* 40. Salisbury: Rural Development Commission.

Brunet, B. (1996) *La Lutte Contre l'Exclusion dans les Territoires Ruraux.* Paris: La Documentation Française.

Bryden, J. M. (ed.) (1994) *Towards Sustainable Rural Communities: The Guelph Seminar Series.* Guelph, Canada: University School of Planning and Development.

Bryden, J. M. (1998) 'Global tendencies and local responses', *LEADER Magazine,* 18: 4–6.

Bryden J.M., Liversedge, A. Robinson, K. and Storey, C. (1998) *Change for the Better: Towards a More Sustainable Rural Development.* Aberdeen: Arkleton Centre for Rural Development Research, University of Aberdeen.

Buller, H. (2000) 'Re-creating rural territories: LEADER in France', *Sociologia Ruralis.* 40(2): 190–9.

Bur, A.-M., Stevens, A. and Young, L. (1999) *Include Us In: Participation for Social Inclusion in Europe.* Canterbury: European Institute of Social Services.

Butcher, H. (1993) 'Introduction, some examples and definitions, in Butcher, H., Glen, A., Henderson, P. and Smith J. (eds), *Community and Public Policy.* London: Pluto Press.

Cabinet Office Performance and Innovation Unit (1999) *Rural Economies.* London: The Stationery Office.

Campagne, P. (1994) *Analysis of the Impact of the Development Project.* LEADER Dossier. Brussels: LEADER Co-ordinating Unit, AEIDL.

Casley, D.J. and Kumar, K. (1987) *Project Monitoring and Evaluation in Agriculture.* New York: World Bank.

Cavazzani, A. and Moseley, M.J. (eds) (2001) *The Practice of Rural Development Partnerships in Europe: 24 Case Studies in Six European Countries.* Soveria Mannelli, Rubbettino, in association with Countryside and Community Research Unit, Cheltenham and Gloucester College of Higher Education.

Centre for Rural Economy (1999) Newsletter. University of Newcastle upon Tyne.

Chanard, A. Blanchard, A., Cavaco, C., Clément, F., Moseley M.J. and Soto, P. (1994) *Diagnosis of the Area and the Mounting of a Development Project.* LEADER Dossier. Brussels: LEADER Co-ordinating Unit, AEIDL.

Cheltenham Observatory (1998) *Directory of Sustainable Rural Initiatives*. Cheltenham: Countryside and Community Research Unit, in association with Forum for the Future.

Cherrett, T. (1999) United Kingdom; a Research Agenda for Partnerships, in E. Westholm, M. Moseley and N. Stenlas (eds), *Local Partnerships and Rural Development in Europe*. Falun, Dalarna Research Institute.

Cherrett, T. (2000) 'The extensive survey in the UK', in Esparcia, J. *et al.* (2000) *op. cit.*: 153–78.

Cherrett, T. (2001) West Tyrone Rural 2000, in A. Cavazzani, and M. Moseley (eds), *The Practice of Rural Development Partnerships in Europe; 24 Case Studies in Europeon Countries*. Soveria Mannelli, Rubbettino.

Cherrett, T. and Moseley, M.J. (2001) *Rural Development Partnerships: Guidelines for Good Practice in the UK*. Cheltenham: Countryside and Community Research Unit, Cheltenham and Gloucester College of Higher Education.

Cloke, P. Milbourne, P. and Thomas, C. (1994) *Lifestyles in Rural England*. Salisbury: Rural Development Commission.

Commissariat Général du Plan (1991) *Guide du Diagnostic pour le Développement Rural*. Paris: Commissariat General du Plan, 5 rue Casimir-Perier 75007.

Commission of the European Communities (1988) *The Future of Rural Society*, COM (88) 371 Final Brussels: Commission of the European Communities.

Commission of the European Communities (1994a) *LEADER II: Innovation in the Service of Rural Society*. Brussels: Commission of the European Communities, Directorate General for Agriculture.

Commission of the European Communities (1994b) A Notice to the Member States ... A Community Initiative for Rural Development. *Official Journal of the European Communities*, C 180, 37(1), July: 48–59.

Committee for Rural Dorset (1994) *Dorset Rural Development Strategy 1994*. Dorchester: Dorset County Council.

Countryside Agency (2000a) *Not Seen, Not Heard? Social Exclusion in Rural Areas*. Cheltenham: The Countryside Agency.

Countryside Agency (2000b) *The State of the Countryside 2000*. Cheltenham: The Countryside Agency.

Countryside Agency (2001a) *The State of the Countryside 2001*. Cheltenham: The Countryside Agency.

Countryside Agency (2001b) *Vital Villages: Equipping Communities to Shape Their Futures*. Cheltenham: The Countryside Agency.

Countryside Agency (2001c) *Market Towns Health Check Handbook*. Cheltenham: The Countryside Agency.

Countryside Agency, English Heritage, English Nature and Environment Agency (2001) *Quality of Life Capital: Managing Environmental, Social and Economic Benefits: An Overview Report*. London: the three Agencies.

Countryside and Community Research Unit (1998) *Village Appraisals for Windows Software Package*. Cheltenham: Cheltenham & Gloucester College of Higher Education and Gloucestershire Rural Community Council.

CRISP (1999) *Newsletter No. 8: September*: www.crisp.org.uk

CRISP (2001) *Get Involved ... It is Your Business: An Introduction to Social and Community Enterprise*. Bristol: CRISP (for copies contact www.sustainable-place.co.uk).

CSR Partnership (2000) *Marches LEADER II Programme: Evaluation and Look Forward – Final Report*. Birmingham: CSR Partnership Ltd.

Cumbria Social Economy Forum (undated) *Community Enterprise: A Self Help Guide*. Penrith: Voluntary Action Cumbria.

Denham, C. and White, I. (1998) 'Differences in urban and rural Britain'. *Population Trends,* 91, Spring: 23–34.

DETR (1997) *Involving Communities in Urban and Rural Regeneration* (2nd edn). London: Department of Environment, Transport and the Regions.

DETR (1999) *Community Enterprise: Good Practice Guide*. London: Department of Environment, Transport and the Regions.

Department of Environment, Transport and the Regions (2001) *Local Strategic Partnerships – Government Guidance Summary* (www.local-regions.dtlr.gov.uk/lsp/guidance/index)

Department of Environment, Transport and the Regions and Ministry of Agriculture, Fisheries and Food (2000) *Our Countryside: The Future* (The 2000 'Rural White Paper'), Cm 4909. London: HMSO.

Derounian, J. G. (1998) *Effective Working with Rural Communities*. Chichester: Packard Publishing.

Development Trusts Association (1998) *Development Trusts: Regenerating Local Communities*. London: the Development Trusts Association.

Dorset Community Action (1998) *Your Place, Then the Planet: A Guide to Creating a Sustainable Place Where you Live*. Dorchester: Dorset Community Action.

Douglas, D.J.A. (1994) 'Strategic planning and management in community economic development', in D.J.A Douglas (ed.) *Community Economic Development in Canada, volume 1*. Whitby, Ontario: McGraw-Hill Ryerson.

Dunn, J. Hodge, I. Monk, S. and Kiddle, C. (1998) *Developing Indicators of Rural Disadvantage*. Salisbury: Rural Development Commission.

Edwards, W., Goodwin, M., Pemberton, S. and Woods, M. (2000) *Partnership Working in Rural Regeneration*. Bristol: The Policy Press and Joseph Rowntree Foundation.

Environment Trust Associates and Local Government Management Board (1994) *Creating Involvement: A Handbook of Tools and Techniques for Effective Community Involvement*. Luton. LGMB.

Esparcia, J., Moseley, M.J. and Noguera, J. (eds) (2000) *Exploring Rural Development Partnerships in Europe: An Analysis of 330 Local Partnerships Across Eight EU Countries*. Valencia: UDERVAL University of Valencia in association with Countryside and Community Research Unit, Cheltenham and Gloucester College of Higher Education.

European Commission (1997) *Territorial Employment Pacts: Examples of Good Practice*. Luxembourg: Office for Official Publications of the European Communities.

Fischler, F. (1998) *Town-Country Europe: Initiatives for Rural Development*. Brussels: Supplement to 'Info LEADER' No. 65, LEADER Observatory, AEIDL.

Flanagan, N., Haase, T., and Walsh, J. (1995) *Planning for Change: A Handbook on Strategic Planning for Local Development Partnerships*. Dublin: Combat Poverty Agency.

Fondation Rurale de Wallonie (1996) *Consultation Villageoise et Développement Rural: Trois Experiences Locales de Consultation de la Population*. Arlon: Fondation Rurale de Wallonie.

Gloucestershire Health (1994) *Health in Gloucestershire: A Strategic Framework for Debate*. Gloucester: Gloucestershire Health.

Goodwin, M. (1998) 'The governance of rural areas: some research issues and agendas', *Journal of Rural Studies,* 14(1): 5–12.

Hambleton, R., Essex, S., Mills, L., Razzaque, K. (1995) *The Collaborative Council: A Study of Inter-Agency Working in Practice.* London: Joseph Rowntree Foundation.

Hansen, G. (1993) 'The promotion and development of entrepreneurial initiatives for employment and enterprise creation', in A.L.V. Alonso and D.N.Short (eds), *The Design and Implementation of Strategies for Local Employment and Economic Development.* Brussels: International Labour Organisation and Commission for the European Communities.

Harvey, B. (1994) *Combating Exclusion: Lessons from the Third EU Poverty Programme in Ireland 1989–94.* Dublin: Combat Poverty Agency.

Kahila, P. (1999) 'Local development policies in rural municipalities', *Finnish Journal of Rural Research and Policy* (special English language edition on 'New Rural Policy'), 7 (2): 75–80.

Keane, M. (1998) 'Rural and local development in Ireland: exploring the theory–practice interface'. *Regional Studies,* 31: 173–7.

Kearney, B., Boyle G.E. and Walsh, J.A. (1994*) EU LEADER Initiative in Ireland: Evaluation and Recommendations*. Dublin: Department of Agriculture, Agriculture House.

Keeble, D.E., Tyler, P., Broom, G., and Lewis, J. (1992) *Business Success in the Countryside: The Performance of Rural Enterprise.* PA Consultants for the Department of Environment. London: HMSO.

LEADER Observatory (1994a) *LEADER Magazine,* Special Issue (no. 7) on 'Innovation in *Constructing Rural Europe'.* Brussels: LEADER Observatory, AEIDL.

LEADER Observatory (1994b) *Support for Small and Medium-Sized Rural Enterprises.* Brussels: LEADER Observatory, AEIDL.

LEADER Observatory (1995a) *Exploiting Local Agricultural Resources: the Experience of LEADER I.* Brussels: LEADER Observatory, AEIDL.

LEADER Observatory (1995b) *Launching and Managing the Local Development Project: the Experience of LEADER I.* Brussels: LEADER Observatory, AEIDL.

LEADER Observatory (1997a) *LEADER Magazine Special Issue (no. 13) on the Cork Conference.* Brussels: LEADER Observatory, AEIDL.

LEADER Observatory (1997b) *Innovation and Rural Development.* Brussels: LEADER Observatory, AEIDL.

LEADER Observatory (1997c) *Evaluating a Territory's Touristic Potential.* Brussels: LEADER Observatory, AEIDL.

LEADER Observatory (1997d) *Organising Local Partnership*. Brussels: LEADER Observatory, AEIDL.

LEADER Observatory (1998) *From Strategy to Action: Project Selection.* Brussels: LEADER Observatory, AEIDL.

LEADER Observatory (1999a) *Territorial Competitiveness: Creating a Territorial Development Strategy in the Light of the LEADER Experience.* Brussels: LEADER Observatory, AEIDL.

LEADER Observatory (1999b) *Information Technologies and Rural Development.* Brussels: LEADER Observatory, AEIDL.

LEADER Observatory (1999c) *Mainstreaming LEADER in Future Rural Policies.* Brussels: LEADER Observatory, AEIDL.

LEADER Observatory (1999d) *Assessing the Value Added of the LEADER Approach*. Brussels: LEADER Observatory, AEIDL.

LEADER Observatory (1999e) *Developing Rural Services*. Brussels: LEADER Observatory, AEIDL.

LEADER Observatory (1999f) *Mainstreaming LEADER in Rural Policies*. Brussels: LEADER Observatory, AEIDL.

LEADER Observatory (1999g) *Innovative Actions in Rural Development: A Directory*. Brussels: LEADER Observatory, AEIDL.

LEADER Observatory (2000a) *Environmental Competitiveness: Creating a Territorial Development Strategy in the Light of the LEADER Experience*. Brussels: LEADER Observatory, AEIDL.

LEADER Observatory (2000b) *Social Competitiveness: Creating a Territorial Development Strategy in the Light of the LEADER Experience*. Brussels: LEADER Observatory, AEIDL.

LEADER Observatory (2000c) *Economic Competitiveness: Creating a Territorial Development Strategy in the Light of the LEADER Experience*. Brussels: LEADER Observatory, AEIDL.

LEADER Observatory (2000d) *Global Competitiveness of Rural Areas: Creating a Territorial Development Strategy in the Light of the LEADER Experience*. Brussels: LEADER Observatory, AEIDL.

LEADER Observatory (2000e) *Training in Aid of Territorial Development*. Brussels: LEADER Observatory, AEIDL.

LEADER Observatory (2000f) *Fighting Social Exclusion in Rural Areas*. Brussels: LEADER Observatory, AEIDL.

LEADER Observatory (2001) *LEADER Magazine*, Special Issue (no. 25) on the European Rural Model. Brussels: LEADER Observatory, AEIDL.

LEADER Observatory (undated a) *Support Systems for New Activities in Rural Areas*. Brussels: LEADER Observatory, AEIDL.

LEADER Observatory (undated b) *Actions Innovantes de Développement Rural*. Brussels: LEADER Observatory, AEIDL.

Liepins, R. (2000) 'New energies for an old idea: reworking approaches to "Community" in contemporary rural studies', *Journal of Rural Studies*, 16: 23–35.

Lowe, P. and Murdoch, J. (1993) *Rural Sustainable Development*. Salisbury: Rural Development Commission.

Lowe, P., Ray, C., Ward, N., Wood, D., Woodward, R. (1998) *Participation in Rural Development: A Review of European Experience*. Centre for Rural Economy Research Report, University of Newcastle upon Tyne.

Lumb R (1990) 'Rural community development: process versus product', in Buller, H. and Wright, S. (eds) *Rural Development: Problems and Practices*. Aldershot: Avebury.

Macdonald, R., Steenberg, C., Harris, P. and Newman, S. (1998) 'Creating a sustainable rural economic development strategy', *Local Governance*, 24(1): 67–76.

Mannion, J. (1996) *Rural Development: Performance and Challenges*. Paper presented to conference 'Rural Development: Striking the Proper Balance'. Co. Limerick: VEC Limerick.

Marsden, T. (1999) 'Rural futures: the consumption countryside and its regulation', *Sociologia Ruralis*, 39(4): 501–20.

Midmore, P. (1998) 'Rural policy reform and local development programmes: appropriate evaluation procedures', *Journal of Agricultural Economics*, 49(3): 409–26.

Midmore, P., Ray, C. and Tregear, A. (1994) *The South Pembrokeshire LEADER Project Evaluation*. University of Wales Aberystwyth, Department of Agricultural Sciences.

Ministère de l'Equipement, des Transports et du Logement (1997) *Construire un projet de territoire: du diagnostic aux strategies*. Paris: Ministére de l'Equipement, DAFU.

Ministry of Agriculture, Fisheries and Food (1994a) *Success with Farm Diversification: A Step by Step Guide*. London: MAFF.

Ministry of Agriculture, Fisheries and Food (1994b) *Success with Marketing Diversified Enterprises*. London: MAFF.

Moseley, M.J. (1979) *Accessibility: The Rural Challenge*. London: Methuen.

Moseley, M.J. (1996a) 'Accessibility and care in a rural area – the case of Tewkesbury Borough', *Research, Policy and Planning*, 14(2): 19–25.

Moseley, M.J. (1996b) 'Baseline studies for local rural development programmes: towards a methodology', *Planning Practice and Research*, 11(1): 19–36.

Moseley, M.J. (1997a) 'Parish appraisals as a tool of rural community development: an assessment of the British experience', *Planning Practice and Research*, 12 (3), 197–212.

Moseley, M.J. (1998) *An Evaluation of the Impact of the Dorset Rural Development Programme*. Cheltenham: Cheltenham and Gloucester College of Higher Education.

Moseley, M.J. (2000a) 'Innovation and rural development: some lessons from Britain and western Europe', *Planning, Practice and Research*, 15 (1–2): 195–215.

Moseley, M.J. (2000b) *An Interim Evaluation of CRISP: The Community Regeneration in the South-West SRB Project*. Unpublished report. Cheltenham: CCRU.

Moseley, M.J. (2000c) 'England's village services in the 1990s: entrepreneurialism, community involvement and the state', *Town Planning Review*, 71(4), 415–33.

Moseley, M.J. (ed.) (2001) *Partnerships for Rural Integrated Development in Europe (The 'PRIDE' Research Project): Final Report to the Research Directorate of the European Commission* (FAIR CT 98-4445). Cheltenham: Countryside and Community Research Unit, Cheltenham and Gloucester College of Higher Education.

Moseley, M.J. (2002) *Village Action Plans: A Research Report*. Cheltenham: The Countryside Agency.

Moseley, M.J. (ed.) (2003) *Local Partnerships for Rural Development; the European Experience*. Wallingford, CABI.

Moseley, M.J., Derounian, J.G. and Allies, P.J. (1996) 'Parish appraisals – a spur to local action? A review of the Gloucestershire and Oxfordshire Experience', *Town Planning Review*, 67 (3): 309–29.

Moseley, M.J. and Cherrett, T. (1993) *Involving People in Local Development*. Brussels: LEADER Observatory, AEIDL.

Moseley, M.J. and Packman, J. (1983) *Mobile Services in Rural Areas*. Norwich: UEA Press.

Moseley, M.J. and Parker, G. (1998) *The Joint Provision of Rural Services*. Salisbury: Rural Development Commission.

Moseley, M.J., Parker, G. and Wragg, A. (2000) *The Joint Provision of Services 2000*. Cheltenham: Countryside Agency.

Murdoch, J. and Abram, S. (1998) 'Defining the limits of community governance', *Journal of Rural Studies,* 14(1): 41–50.

Murray, M. and Greer, J. (2001) *Participatory Village Planning: Practice Guidelines Workbook*. Belfast: Rural Innovation and Research Partnership, Queen's University.

National Economic and Social Council (1994) *New Approaches to Rural Development.* Dublin, NESC.

National Rural Enterprise Centre (1998) *NREC's Vision for the Future of Rural Services: A Discussion Paper*. Stoneleigh: National Rural Enterprise Centre.

NCH (undated) *Challenging the Rural Idyll.* Cheltenham: The Countryside Agency.

New Economics Foundation (1998) *Participation Works! Twenty-one Participatory Techniques for the Twenty First Century*. London: New Economics Foundation.

New Economics Foundation (2001) *Plugging the Leaks – Newsletters 1,2,3.* London: New Economics Foundation.

New Economics Foundation (undated) *Community Works! A Guide to Community Economic Action.* London: New Economics Foundation.

Newby, H. (1985) *Green and Pleasant Land?* Harmondsworth: Penguin.

Norberge-Hodge, H. (1999) 'Bringing the economy back home: towards a culture of place', *The Ecologist,* 29(3): 215–8.

North, D. and Smallbone, D. (1993) *Small Business in Rural Areas.* Salisbury: Rural Development Commission.

Northamptonshire ACRE (1999) *Improving Your Village: Community Information Pack.* Northampton: Northamptonshire ACRE.

O'Cinneide, M. and Cuddy, M. (eds) (1992) *Perspectives on Rural Development in Advanced Economies.* Galway: Centre for Development Studies, University College.

OECD (1990) *Rural Development Policy.* Paris: Organisation for Economic Co-operation and Development.

OECD (1991) *New Ways of Managing Services in Rural Areas.* Paris: Organisation for Economic Co-operation and Development.

OECD (1996) *Ireland: Local Partnerships and Social Innovation.* Paris: Organisation for Economic Co-operation and Development.

Owen, S. (2002) 'Locality and community: towards a vehicle for community-based decision making in rural localities in England', *Town Planning Review,* 73 (1): 1–21.

Peak District Rural Deprivation Forum (undated) *Uphill Struggles: A Report of the Conference on the Evaluation of the Peak District Rural Deprivation Forum.* Hope: The Peak District Rural Deprivation Forum.

PIEDA Consulting (1997) *Interim Evaluation of the English LEADER II Programme.* Manchester: PIEDA Consulting.

Policy Studies Institute (1998) *Rural Disadvantage – Understanding the Processes.* London: Rural Development Commission.

Ray, C. (1996a) *The Dialectic of Local Development: the Case of the EU LEADER I Rural Development Programme.* Working Paper 23. Newcastle upon Tyne: Centre for Rural Economy, University of Newcastle upon Tyne.

Ray, C. (1996b) 'Local rural development and the LEADER I Programme', in P. Allanson and M. Whitby (eds) *The Rural Economy and the British Countryside.* London: Earthscan.

Ray, C. (1997) *Culture Economies and Local Development.* Working Paper 25. University of Newcastle Upon Tyne, Newcastle Upon Tyne: Center for Rural Economy.

Ray, C. (1998) *New Places and Space for Rural Development in the European Union: An Analysis of the UK LEADER I Programme.* Working Paper 34. Newcastle upon Tyne: University of Newcastle upon Tyne, Centre for Rural Economy.

Ray, C. (2000a) 'The EU LEADER Programme: Rural Development Laboratory', *Sociologia Ruralis,* 40 (2): 165–71.

Ray, C. (2000b) 'Endogenous socio-economic development in the European Union: issues of evaluation', *Journal of Rural Studies,* 16: 447–58.

Ray, C. (2001) *Culture Economies: A Perspective on Local Rural Economies in Europe.* Newcastle upon Tyne: Centre for Rural Economy, University of Newcastle upon Tyne.

Rogers, A.W. (1993) *English Rural Communities: An Assessment and Prospects for the 1990.* Salisbury: Rural Development Commission.

Rogers, E. M. (1995) *Diffusion of Innovations* (4th edn). New York: The Free Press.

(RSPB) Royal Society for the Protection of Birds, Countryside Agency and Cheltenham and Gloucester College of Higher Education (1999) *Working Together: Communities, Conservation and Rural Economies.* Sandy: RSPB.

Rural Development Commission (1993) *The Economy of Rural England.* Salisbury: Rural Development Commission.

Rural Development Commission (1994) *Guide to Preparing Rural Development Programme Strategies.* Salisbury: Rural Development Commission.

Rural Development Commission (1996a) *Country Lifelines: Good Practice in Rural Transport.* Salisbury: Rural Development Commission.

Rural Development Commission (1999) *Rural Housing.* Salisbury: Rural Development Commission.

Rural Development Council for Northern Ireland (undated) *Developing Rural Enterprise: Providing Specialist Support to the Rural Enterprise Network.* Magherafelt, Northern Ireland: Rural Development Council.

Rural Forum and the Scottish Office (1997) *Community Involvement in Rural Development Initiatives.* Good Practice in Rural Development no. 2. Edinburgh: Scottish Office Central Research Unit.

Ruttan, V.W. (1996) 'What happened to technology adoption-diffusion research?', *Sociologia Ruralis,* 36 (1): 51–73.

Scott, D. and Russell, L. (1999) *Uphill Struggles: An Evaluation of the Rural Deprivation Forum in the Peak District.* Hope: The Peak District Deprivation Forum.

Scott, D. Shenton, N. and Healey, B. (1991) *Hidden Deprivation in the Countryside: Local Studies in the Peak National Park.* Glossop: The Peak Park Trust.

Scottish National Rural Partnership (1998) *New Ideas in Rural Development No. 6: Becoming an Entrepreneur in Rural Scotland.* Edinburgh: Scottish Office Central Research Unit.

Segal Quince Wicksteed (1999) *Programme Evaluation of LEADER II in the Eastern Region and the Development of Forward Strategies.* Cambridge: Segal Quince Wicksteed.

Selman, P. (1996) *Local Sustainability: Managing and Planning Ecologically Sound Places.* London: Paul Chapman Publishing.

Shucksmith, M. (2000) *Exclusive Countryside? Social Inclusion and Regeneration in Rural Areas*. York: Joseph Rowntree Foundation.

Shucksmith, M., Chapman, P. and Clark, G. (1994) *Disadvantage in Rural Scotland: A Summary Report*. Perth: Rural Forum.

Shucksmith, M., Roberts, D., Scott, D., Chapman, P., and Conway, E. (1996) *Disadvantage in Rural Areas*. Salisbury: Rural Development Commission.

Slee, W. and Snowden, P. (1997) *Effective Partnership Working*. Good Practice in Rural Development No. 1. Edinburgh: Scottish National Rural Partnership, Scottish Office.

*Sociologia Ruralis* (2000) Special Issue on 'LEADER', 40(2).

South West Social Economy Partnership (2000) *Supporting the Social Economy: A Development Plan for the South-West*. SWSEP (no place of publication indicated).

Stoker, G. (1996) 'The struggle to reform local government: 1970–95', *Public Money and Management*, 16 (1): 17–22.

Taylor, R. (1998) 'Rural action for the environment: a review', in Derounian, J.G. *Effective Working with Rural Communities*. Chichester: Packard Publishing.

UK LEADER II Network (2000) *Making a Difference in Rural Areas*. London: LRDP.

Vuarin, P. and Rodriguez, M. (1994) 'Innovation and communication within LEADER', *LEADER Magazine*, 7: 13–16.

Walsh, J. (1995) *Local Development Theory and Practice: Recent Experience in Ireland*. Conference on 'Sustainable Regional and Local Development', Maynooth, Ireland.

Walsh, K. (2000) *The Potential of Participatory Evaluation for Rural Community Development*. Unpublished PhD thesis, University of Bristol.

Warburton, D. (1998a) *Participatory Action in the Countryside: A Literature Review*. Cheltenham: Countryside Commission.

Warburton, D (1998b) *The Achievements and Effectiveness of 'Rural Action': An Evaluation*. Report to the Countryside Commission. Cheltenham: Countryside Commission.

Westholm, E., Moseley, M.J. and Stenlas, N. (eds) (1999) *Local Partnerships and Rural Development in Europe: a Literature Review of Practice and Theory*. Sweden: Dalarna Institute, in association with CCRU, Cheltenham & Gloucester College of Higher Education.

Winter, M. (1997) 'New policies and skills: agricultural change and technology transfer', *Sociologia Ruralis* 37(3): 363–81.

# Index